Springer Undergraduate Texts
in Mathematics and Technology

Alexander Shen

# Algorithms and Programming

## Problems and Solutions

### Second Edition

 Springer

Alexander Shen
Laboratoire d'Informatique Fondamentale de Marseille (LIF)
CNRS, Université de la Méditerranée, Université de Provence
CMI 39 Rue Joliot-Curie
13453 Marseille Cedex 13
France
alexander.shen@lif.univ-mrs.fr

and

Russian Academy of Sciences
Institute for Information Transmission Problems
Bolshoy Karetny per. 19
Moscow, GSP-4, 127994
Russia

*Series Editors*

Jonathan M. Borwein, FRSC
Professor Laureate
Director Centre for
  Computer Assisted Research Mathematics
  and its Applications, CARMA
School of Mathematical & Physical Sciences
University of Newcastle
Callaghan NSW 2308
Australia
Jonathan.Borwein@newcastle.edu.au

Helge Holden
Department of Mathematical Sciences
Norwegian University of Science and
  Technology
Alfred Getz vei 1
NO-7491 Trondheim
Norway
holden@math.ntnu.no

ISSN 1867-5506                 e-ISSN 1867-5514
ISBN 978-1-4939-3700-4         ISBN 978-1-4419-1748-5  (eBook)
DOI 10.1007/978-1-4419-1748-5
Springer New York Dordrecht Heidelberg London

Mathematics Subject Classification (2000): 65K05, 65Yxx, 90Cxx, 68-01, 68W40

1st edition: © Birkhäuser 1997
Reprint of 1st edition in series: 'Modern Birkhäuser Classics' © Birkhäuser 2008

2nd edition: © Springer Science+Business Media, LLC 2010
Softcover re-print of the Hardcover 2nd edition 2010

Printed on acid-free paper

Springer is part of Springer Science+Business Media (www.springer.com)

*To the memory of Anna Pogossiants*

# Contents

# Preface to the Second Edition

Somebody once said that one may prove the correctness of an algorithm, but not of a program. One of the main goals of this book is to convince the reader that things are not so bad.

A well-known programmer, C.A.R. Hoare, said that the beauty of a program is not an additional benefit but a criterion that separates success from failure. If, while solving problems in this book, you come to appreciate the beauty of a well-written program with each part in its correct place, the author's goal will have been reached.

Theoretically this book can be used to study programming without a computer: one could write (correct) programs with pencil and paper. But in practice the ability to run the programs is a challenge and a reward that makes programming a fun.

We have utilized the problem-solution format. Some chapters are collections of problems having a common topic, while others are devoted to one specific algorithm (e.g., chapter 16 covers LR(1)-parsing). The chapters are more or less independent, but the concluding chapters are more difficult. Chapters 1–7 cover material usually included in undergraduate courses while chapters 15–16 are more appropriate for a graduate compiler course. In each chapter we have tried to give problems at different levels starting with easy exercises.

Problems are usually provided with solutions, answers or hints. However, we strongly recommend to read the solution only after the reader makes a good faith attempt to solve it independently.

The book is restricted to "micro-programming" leaving aside another very important topic: how to split the program into a manageable parts with nice interfaces between them. (Probably this can be learned only by reading and modifying rather large programs.)

Pascal is used as a programming language; though being outdated, it is reasonably clear, so the readers familiar with any other procedural language (C, Modula, Oberon, etc.) will encounter no difficulties. For the reader's convenience, a short appendix is added that lists basic differences between Pascal and C. It is intended to help the reader who knows C to understand the program notation in the book (but cannot replace textbooks on C).

Most of the problems, of course, are well known. References are rare, but absence of references does not mean that the problem or algorithm is new. However, we hope that in some cases the algorithm or the proof is explained better than what is found in other sources.

This book is addressed both to the ambitious student who wants to test and improve his/her skills and to the instructor looking for problems for his/her class.

I thank all the people I met while teaching programming (first of all, my former students from 57th school and A.G. Kushnirenko, who was my programming teacher) and all readers that sent me corrections for the preliminary versions of this book (especially Yu.V. Matijasevich).

I also thank American Mathematical Society (former Soviet Union aid fund), International Science Foundation, Open Society Foundation, MIT, University of Bordeaux, Bonn University, the Rosenbaum Foundation, INTAS, University of Provence, CNRS, Institute of Problems of Information Transmission and even the Russian government for support during writing this book.

I thank Ann Kostant, Elizabeth Loew and the other nice people at Birkhäuser and Springer for their help. Tom Scavo did a great job correcting my English (as well as several other errors) but in no case should he be blamed for the remaining mistakes. Peter Panov helped to prepare the second edition by translating new material in chapters 10–12 and language editing.

The Russian version of this book is freely distributable as a TEX source and camera-ready copy; please look at `ftp://ftp.mccme.ru/users/shen/progbook` and/or contact the author (e-mail addresses: `shen@landau.ac.ru`, `shen@mccme.ru`, `sasha.shen@gmail.com`, `alexander.shen@lif.univ-mrs.fr`) for details. I'd be grateful if bug reports could be sent to the same addresses.

*Alexander Shen*
September, 2009

# 1

# Variables, expressions, assignments

In this chapter we begin (section 1.1) with simple programming problems using variables, assignments and basic constructs (if- and while-statements). Then (section 1.2) we introduce arrays and programming techniques related to them. Finally (section 1.3), we consider a useful approach that helps to develop one-pass algorithms; each of the elements of the input array is processed once.

## 1.1 Problems without arrays

**1.1.1.** Consider two integer variables a and b. Write a program block that exchanges the values of a and b (i.e., the value of a becomes the value of b and vice versa).

*Solution.* We use an auxiliary integer variable t.

```
t := a;
a := b;
b := t;
```

If we try to eliminate this auxiliary variable by writing

```
a := b;
b := a;
```

we get an incorrect program (the value of a is lost after the first assignment).

**1.1.2.** Solve the preceding problem without an auxiliary variable. (Assume all variables accept arbitrary integer values.)

*Solution.* (By a0 and b0 we denote the initial values of a and b.)

```
a := a + b; {a = a0 + b0, b = b0}
b := a - b; {a = a0 + b0, b = a0}
a := a - b; {a = b0, b = a0}
```

A. Shen, *Algorithms and Programming*, Springer Undergraduate Texts in Mathematics and Technology, DOI 10.1007/978-1-4419-1748-5_1, © Springer Science+Business Media, LLC 2010

**1.1.3.** Let a be an integer and n be a nonnegative integer. Compute $a^n$. In other words, we ask for a program that does not change the values of a and n and assigns the value $a^n$ to another variable (say, b). (The program may use other variables as well.)

*Solution.* Consider an integer variable k, whose range is 0..n. (We maintain the property: $b = a^k$.)

```
k := 0; b := 1;
{b = a^k}
while k <> n do begin
    k := k + 1;
    b := b * a;
end;
```

Another solution:

```
k := n; b := 1;
{a^n = b * (a^k)}
while k <> 0 do begin
    k := k - 1;
    b := b * a;
end;
```

**1.1.4.** Solve the preceding problem with the additional requirement that the number of execution steps should be of order log n (i.e., it should not exceed $C$ log n for some constant $C$).

*Solution.* Let us make some changes in the second solution of the preceding problem:

```
k := n;  b := 1;  c := a;
{a^n = b * (c^k)}
while k <> 0 do begin
    if k mod 2 = 0 then begin
        k := k div 2;
        c := c*c;
    end else begin
        k := k - 1;
        b := b * c;
    end;
end;
```

In both cases (even k and odd k) the value of k decreases; if k is even, it is divided by 2; if k is odd, after k := k - 1 it becomes even and is divided by 2 during the next iteration. Therefore, after any two iterations k becomes twice smaller (or even less), so the number of steps is logarithmic.                    □

**1.1.5.** Two nonnegative integers a and b are given. Compute the product a*b (only +, −, =, <> are allowed).

*Solution.*

```
k := 0; c := 0;
{invariant relation: c = a * k}
while k <> b do begin
  k := k + 1;
  c := c + a;
end;
{c = a * k and k = b, therefore, c = a * b}
```

**1.1.6.** Two nonnegative integers a and b are given. Compute $a + b$. Only assignments of the form

$$\langle \text{variable1} \rangle := \langle \text{variable2} \rangle;$$

$$\langle \text{variable} \rangle := \langle \text{number} \rangle;$$

$$\langle \text{variable1} \rangle := \langle \text{variable2} \rangle + 1;$$

are allowed.

[*Hint.* Use the invariant relation c=a+k.]

**1.1.7.** A nonnegative integer a and positive integer d are given. Compute the quotient q and the remainder r when a is divided by d. Do not use the operations div or mod.

*Solution.* By definition, $a = q*d + r$ and $0 \leqslant r < d$.

```
{a >= 0; d > 0}
r := a; q := 0;
{invariant relation: a = q*d+r, 0 <= r}
while not (r < d) do begin
  {r >= d}
  r := r - d; {r >= 0}
  q := q + 1;
end;
```

**1.1.8.** For a given nonnegative integer n, compute n! (n! is the product $1 \cdot 2 \cdot 3 \cdots n$; we assume that $0! = 1$).

**1.1.9.** The Fibonacci sequence is defined as follows: $a_0 = 0$, $a_1 = 1$, $a_k = a_{k-1} + a_{k-2}$ for $k \geqslant 2$. For a given n, compute $a_n$.

**1.1.10.** Repeat the preceding problem with the additional requirement that the number of operations should be proportional to $\log n$. (Use only integer variables.)

[*Hint.* Any pair of consecutive Fibonacci numbers is the product of the matrix

$$\begin{bmatrix} 1 & 1 \\ 1 & 0 \end{bmatrix}$$

and the preceding pair. Therefore, it is enough to compute the n-th power of this matrix. It can be done in $C \log n$ steps in the same manner as problem 1.1.4 (for integers).]                                                                                             □

**1.1.11.** For a nonnegative integer n, compute

$$\frac{1}{0!} + \frac{1}{1!} + \cdots + \frac{1}{n!}.$$                                          □

**1.1.12.** Repeat the preceding problem with the additional requirement that the number of steps (i.e., the number of assignments performed during the execution) should be of order n (i.e., not greater than $Cn$ for some constant $C$).

*Solution.* The invariant relation: $\mathrm{sum} = 1/1! + \cdots + 1/k!$, $\mathrm{last} = 1/k!$ (it is important not to compute k! each time from scratch).                                   □

**1.1.13.** Two nonnegative integers a and b are not both zero. Compute GCD(a,b), the greatest common divisor of a and b.

*Solution.* (Version 1)

```
if a < b then begin
| k := a;
end else begin
| k := b;
end;
{k = min (a,b)}
{invariant relation: no numbers greater than k (and
  therefore than a or b) are common divisors}
while not ((a mod k = 0) and (b mod k = 0)) do begin
| k := k - 1;
end;
{k is a common divisor, all larger k are not}
```

(Version 2 — Euclid's algorithm.) We assume that GCD(0,0)=0. Then GCD(a,b) = GCD(a-b,b) = GCD(a,b-a) with GCD(a,0) = GCD(0,a) = a for all a, b $\geqslant$ 0. This property allows us to decrease a and b without changing GCD(a,b).

```
m := a; n := b;
{invariant relation: GCD(a,b) = GCD(m,n); m,n >= 0 }
while not ((m=0) or (n=0)) do begin
    if m >= n then begin
    | m := m - n;
    end else begin
    | n := n - m;
    end;
```

```
end;
{m = 0 or n = 0}
if m = 0 then begin
| k := n;
end else begin {n = 0}
| k := m;
end;                                                    □
```

**1.1.14.** Write down a modified version of Euclid's algorithm that uses the identities

$$GCD(a,b) = GCD(a \bmod b, b) \text{ for } a \geqslant b;$$

$$GCD(a,b) = GCD(a, b \bmod a) \text{ for } b \geqslant a. \qquad \square$$

**1.1.15.** Nonnegative integers a and b are given, at least one of which is not zero. Find $d = GCD(a,b)$ and integers x and y such that $d = a*x + b*y$.

*Solution.* Add the auxiliary variables p, q, r, s to Euclid's algorithm and add the requirements $m = p*a+q*b$ and $n = r*a+s*b$ to the invariant relation:

```
m:=a; n:=b; p:=1; q:=0; r:=0; s:=1;
{invariant relation:
    GCD(a,b) = GCD(m,n);
    m,n >= 0
    m = p*a + q*b;
    n = r*a + s*b.}
while not ((m=0) or (n=0)) do begin
  | if m >= n then begin
  |   | m := m - n;
  |   | p := p - r;
  |   | q := q - s;
  | end else begin
  |   | n := n - m;
  |   | r := r - p;
  |   | s := s - q;
  | end;
end;
if m = 0 then begin
| k := n; x := r; y := s;
end else begin
| k := m; x := p; y := q;
end;                                                    □
```

**1.1.16.** Solve the preceding problem using the mod operator.     □

**1.1.17.** (E. Dijkstra) Let us add three variables u, v, z to Euclid's algorithm:

```
m := a; n := b; u := b; v := a;
{invariant relation: GCD(a,b) = GCD(m,n); m,n>=0}
while not ((m=0) or (n=0)) do begin
  if m >= n then begin
  | m := m - n; v := v + u;
  end else begin
  | n := n - m; u := u + v;
  end;
end;
if m = 0 then begin
| z:= v;
end else begin {n=0}
| z:= u;
end;
```

Prove that after execution the value of z is twice as large as the least common multiple of a and b; that is, $z = 2 \cdot LCM(a,b)$.

*Solution.* Look at the value of $m \cdot u + n \cdot v$, which remains unchanged during program execution. Initially it is equal to 2ab; therefore, this expression has the same value at the end. Now apply the identity $GCD(a, b) \cdot LCM(a, b) = ab$.   □

**1.1.18.** Write a version of Euclid's algorithm using the identities

$$GCD(2a, 2b) = 2 \cdot GCD(a, b); \quad GCD(2a, b) = GCD(a, b) \quad \text{for odd b}$$

The algorithm should avoid division (div and mod operations); only division by 2 and the test "to be even" are allowed. (The number of operations should be of order log k if both numbers do not exceed k.)

*Solution.*

```
m:=a; n:=b; d:=1;
{GCD(a,b) = d * GCD(m,n)}
while not ((m=0) or (n=0)) do begin
  if (m mod 2 = 0) and (n mod 2 = 0) then begin
  | d:= d*2; m:= m div 2; n:= n div 2;
  end else if (m mod 2 = 0) and (n mod 2 = 1) then begin
  | m:= m div 2;
  end else if(m mod 2 = 1) and (n mod 2 = 0) then begin
  | n:= n div 2;
  end else if (m mod 2=1) and (n mod 2=1) then begin
    if m >=n then begin
    | m:= m-n;
    end else begin {m < n}
    | n:= n-m;
    end;
  end;
```

```
end;
{m=0 => answer=d*n; n=0 => answer=d*m}
```

If both numbers m and n do not exceed k, the number of operations does not exceed $C \log k$; indeed, each other operation makes at least one of the numbers m and n twice smaller. □

**1.1.19.** Modify the solution of the preceding problem to find x and y such that $ax + by = GCD(a, b)$.

*Solution.* (The idea was communicated by D. Zvonkin.) Assume that both a and b are even. In this case we divide both of them by 2; the values of x and y we are looking for remain unchanged. Therefore, without loss of generality, we may assume that at least one of the numbers a and b is odd. (This property will remain true.)

As before, we wish to maintain the numbers p, q, r, s such that

$$m = ap + bq$$
$$n = ar + bs$$

The problem, however, is that if we divide m by 2 (say), then we should at the same time divide p and q by 2. In this case p and q are no longer integers but become finite binary fractions; that is, numbers of the type $r/2^s$. Such a number can be represented by a pair $\langle r, s \rangle$. As a result, we get d as a linear combination of a and b with coefficients being finite binary fractions. In other words, we have

$$2^i d = ax + by$$

for some integers x, y and nonnegative integer i. What should we do if $i > 0$? If both x and y are even, we may divide them by 2 (and decrease i by 1). If not, we apply the transformations:

$$x := x + b$$
$$y := y - a$$

(this transformation leaves $ax + by$ unchanged). Let us see why this works. Recall that one of the numbers a and b is odd (according to our assumption). Let a be odd. If y is even, then x is even as well (otherwise $ax + by$ is odd); this case is considered above. If a and y are odd, then y becomes even after executing the statement $y := y - a$. □

**1.1.20.** Write a program that prints the squares of the natural numbers $0, \ldots, n$ for a given $n \geqslant 0$.

*Solution.*

```
k:=0;
writeln (k*k);
{invariant relation: k<=n, all the squares
   up to (k*k) are printed}
```

```
while not (k=n) do begin
  k := k + 1;
  writeln (k*k);
end;                                                        □
```

**1.1.21.** Repeat the preceding problem, but only addition and subtraction are allowed. The number of steps should be of order n.

*Solution.* We use the variable k_square, and maintain the invariant relation k_square = $k^2$:

```
k := 0; k_square := 0;
writeln (k_square);
while not (k = n) do begin
  k := k + 1;
  {k_square = (k-1) * (k-1) = k*k - 2*k + 1}
  k_square := k_square + k + k - 1;
  writeln (k_square);
end;                                                        □
```

*Remark.* We can avoid subtraction by the following trick:

```
while not (k = n) do begin
  k_square := k_square + k;
  {k_square = k*k + k}
  k := k + 1;
  {k_square = (k-1)*(k-1)+(k-1)=k*k-k}
  k_square := k_square + k;
end;
```

**1.1.22.** Write a program that prints the factorization of a given integer n > 0. (In other words, it should print prime numbers whose product is equal to n; if n = 1, nothing should be printed.)

*Solution.* (Version 1)

```
k := n;
{invariant relation: the product of k and all numbers
 printed is equal to n; only prime numbers are printed}
while not (k = 1) do begin
  t := 2;
  {invariant relation: k has no divisors in (1,t)}
  while k mod t <> 0 do begin
    t := t + 1;
  end;
  {t is the smallest divisor of k greater than 1;
   therefore, t is prime}
  writeln (t);
```

```
  | k:=k div t;
  end;
```

(Version 2)

```
    k := n; t := 2;
    {the product of k and all number printed is equal
     to n; only prime numbers are printed;
     k has no divisors in (1,t)}
    while not (k = 1) do begin
      | if k mod t = 0   then begin
      |   | {k is a multiple of t and has no divisors
      |   |    less than t; therefore, t is prime}
      |   | k := k div t;
      |   | writeln (t);
      | end else begin
      |   | {k is not a multiple of t}
      |   | t := t+1;
      | end;
    end;
```
□

**1.1.23.** Solve the preceding problem taking into account the following fact: any composite number $N$ has a factor not exceeding $\sqrt{N}$.

*Solution.* In version 2 of the above solution, replace t:=t+1 by

```
    if t*t > k then begin
    | t:=k;
    end else begin
    | t:=t+1;
    end;
```
□

**1.1.24.** Check whether a given number n > 1 is prime. □

**1.1.25.** (This problem requires some algebra) A Gaussian integer $n + mi \in \mathbb{Z}[i]$ is given. (a) Check whether it is a prime element in $\mathbb{Z}[i]$; (b) print its factorization as a product of prime factors in $\mathbb{Z}[i]$. □

**1.1.26.** Assume the command write(i) is allowed for $i = 0, 1, 2, \ldots, 9$. Write a program that prints the decimal representation of a given positive integer n.

*Solution.*

```
    base:=1;
    {base is an integer power of 10 not exceeding n}
    while 10 * base <= n do begin
    | base:= base * 10;
    end;
```

```
{base is a maximal power of 10 not exceeding n}
k:=n;
{invariant relation: it remains to print k with the
 same number of digits as in base; base = 100..00}
while base <> 1 do begin
  write(k div base);
  k:= k mod base;
  base:= base div 10;
end;
{base=1; it remains to write one digit k}
write(k);
```

Please note that this program assumes that n > 0.                    □

(A typical mistake while solving this problem is that the numbers with zeros in the middle are printed incorrectly. The invariant relation mentioned above allows the case k < base; in this case, the decimal representation of k begins with zero.)

**1.1.27.** Write a program that prints the decimal representation of a positive integer n in reverse. (For n = 173, the program should print 371.)

*Solution.*

```
k:= n;
{invariant relation: it remains to print k reversed}
while k <> 0 do begin
  write (k mod 10);
  k:= k div 10;
end;
```
                                                                     □

**1.1.28.** A nonnegative integer n is given. Count all the solutions of the inequality $x^2 + y^2 < n$ where x and y are nonnegative integers. The program should not use operations with real numbers (square roots, etc.).

*Solution.*

```
a := 0; s := 0;
{invariant relation: s = the number of all pairs
 <x,y> such that (x*x + y*y < n and x < a)}
while a*a < n do begin
  ...
  {t = the number of nonnegative integers y such that
   a*a + y*y < n (for fixed a)}
  a := a + 1;
  s := s + t;
end;
{a*a >= n, therefore s is the total number of solutions}
```

The ellipsis represents part of the program that is still to be written. Here it is:

```
b := 0; t:= 0;
{invariant relation: t is the number of integers y
   such that (a*a + y*y < n and 0<=y<b)}
while a*a + b*b < n do begin
| b := b + 1;
| t := t + 1;
end;
{a*a + b*b >= n,  so t is the number of nonnegative
   integers y such that a*a + y*y < n}
```
□

**1.1.29.** The same problem with the additional restriction that the total number of operations should be of order $\sqrt{n}$. (The previous solution requires about n operations.)

*Solution.* We have to count all the integer grid points in the first quadrant that lie inside the circle of radius $\sqrt{n}$. The set in question (call it X) is a union of columns of points having width 1 and non-increasing height.

The idea is to trace the boundary of this set, which resembles a staircase that goes down as we move from left to right. The current position is <a,b>. We use one more variable s and maintain the following invariant relation:

<a,b> is on the top of a-th column;
s is the number of points in the preceding columns.

Formally,

- b is minimal among all $b \geqslant 0$ such that <a,b> is not in X;
- s is the number of all pairs <x,y> of nonnegative integers such that x < a and <x,y> ∈ X.

These conditions will be denoted by (I).

```
a := 0; b := 0;
while <0,b> is in X do begin
| b := b + 1;
end;
{a = 0, b is minimal among all b >= 0
   such that <a,b> is not in X }
s := 0;
{invariant relation: (I)}
while not (b = 0) do begin
```

```
  s := s + b;
  {s is the number of points in columns 0..a}
  a := a + 1;
  {point <a,b> is outside X, it should be moved down
      to restore (I) unless (I) is already true}
  while (b <> 0) and (<a, b-1> is not in X) do begin
  | b := b - 1;
  end;
end;
{(I), b = 0, therefore the a-th column and all
  subsequent columns are empty; s is the required number}
```

An estimate for the number of steps is evident. First we move up performing not more than $\sqrt{n}$ steps. Then we move right and down in not more than $\sqrt{n}$ steps in each direction.                                                                          □

**1.1.30.** Nonnegative integers n and k are given, with n > 1. Print k digits of the decimal representation of the number $1/n$. (If two decimal representations exist, such as $0.499\ldots = 0.500\ldots$, print the latter.) The program should use integer variables only.

*Solution.* Moving the decimal point of the number $1/n$, k positions to the right, we get the number $10^k/n$. We wish to print its integer part; that is, we must compute $10^k$ div n. We do not want to compute $10^k$ because of the possibility of integer overflow. Instead, we perform ordinary division. Here is the program:

```
  m := 0;
  r := 1;
  {m digits of 1/n  are printed; it remains to print
    k - m digits of the decimal expansion of r/n}
  while m <> k do begin
  | write ( (10 * r) div n);
  | r := (10 * r) mod n;
  | m := m + 1;
  end;
```
                                                                          □

**1.1.31.** A natural number n > 1 is given. Find the length of the period of the decimal number $1/n$.

*Solution.* The period of a decimal fraction is equal to the period of the sequence of "remainders" r (see the solution of the preceding problem). [Prove this fact: do not forget to prove that the period of the fraction cannot be less than the period of the sequence of remainders.] In the sequence of remainders all terms that form the period are distinct and the length of the non-periodic initial segment does not exceed n. Therefore, it is enough to find the $(n + 1)$-th term of the sequence, and then to find the minimal k such that the $(n + 1 + k)$-th term is equal to the $(n + 1)$-th term.

```
m := 0;
r := 1;
{r/n = what remains from 1/n after the decimal point
    is moved m positions to the right and the integral
    part is discarded}
while m <> n+1 do begin
│ r := (10 * r) mod n;
│ m := m + 1;
end;
c := r;
{c = (n+1)-th term of the sequence of remainders}
r := (10 * r) mod n;
k := 1;
{r = (n+k+1)-th term of the same sequence}
while r <> c do begin
│ r := (10 * r) mod n;
│ k := k + 1;
end;                                                        □
```

**1.1.32.** (R.W. Floyd, communicated by Yu.V. Matijasevich) A function f that maps {1..N} to {1..N} is given. The sequence 1, f(1), f(f(1)), ... is periodic. Find its period. The number of operations should be proportional to the length of the smallest initial segment that includes the period (this length may be significantly less than N).

*Solution*. After discarding the initial segment, we have a periodic sequence, and all terms in the period are different.

```
{Notation: f[n,1]=f(f(...f(1)...)) (n times)}
k := 1;
a := f(1);
b := f(f(1));
{a = f[k,1]; b = f[2k,1]}
while a <> b do begin
│ k:=k+1; a:=f(a); b:=f(f(b));
end;
{a = f[k,1] = f[2k,1]; f[k,1] is in the periodic part}
m := 1;  b := f(a);
{b = f[k+m,1]; f[k,1],...,f[k+m-1,1] are different}
while a <> b do begin
│ m:=m+1; b:=f(b);
end;
{period = m}                                                □
```

Note that the value of *k* obtained after the first loop may be greater than the actual period.

**1.1.33.** (E. Dijkstra). A function f whose arguments and values are nonnegative integers is defined as follows: $f(0) = 0$, $f(1) = 1$, $f(2n) = f(n)$, $f(2n + 1) = f(n) + f(n + 1)$. Write a program that computes $f(n)$ for a given n; the number of operations should be of order $\log n$.

*Solution.*

```
k := n; a := 1; b := 0;
{invariant relation: 0<=k, f(n) = a*f(k) + b*f(k+1)}
while k <> 0 do begin
  if k mod 2 = 0  then begin
    l := k div 2;
    {k=2l, f(k)=f(l), f(k+1) = f(2l+1) = f(l) + f(l+1),
     f (n) = a*f(k) + b*f(k+1) = (a+b)*f(l) + b*f(l+1)}
    a := a + b; k := l;
  end else begin
    l := k div 2;
    {k = 2l + 1, f(k) = f(l) + f(l+1),
     f(k+1) = f(2l+2) = f(l+1),
     f(n) = a*f(k) + b*f(k+1) = a*f(l) + (a+b)*f(l+1)}
    b := a + b; k := l;
  end;
end;
{k = 0, f(n) = a * f(0) + b * f(1) = b,
 b is the answer}
```

**1.1.34.** The same problem for $f(0) = 13$, $f(1) = 17$, $f(2) = 20$, $f(3) = 30$, $f(2n) = 43\,f(n) + 57\,f(n + 1)$ and $f(2n + 1) = 91\,f(n) + 179\,f(n + 1)$ for $n \geqslant 2$.

[*Hint.* The program stores k, a, b, c such that $f(n) = a \cdot f(k) + b \cdot f(k+1) + c \cdot f(k+2)$.]

**1.1.35.** Two nonnegative integers a and b are given, with $b > 0$. Find a `mod` b and a `div` b using only integer variables and avoiding explicit `div` and `mod` operations (the only exception: an even number may be divided by 2). The number of operations should not exceed $C_1 \log(a/b) + C_2$ for some constants $C_1$ and $C_2$.

*Solution.*

```
b1 := b;
while b1 <= a do begin
  b1 := b1 * 2;
end;
{b1 > a, b1 = b * (integer power of 2)}
q:=0; r:=a;
{invariant relation: q, r are quotient and remainder when
  a is divided by b1; b1 = b * (some integer power of 2)}
  while b1 <> b do begin
```

```
    b1 := b1 div 2 ; q := q * 2;
    { a = b1 * q + r, 0 <= r < 2 * b1}
    if r >= b1 then begin
      r := r - b1;
      q := q + 1;
    end;
end;
{q, r are quotient and remainder when a is divided by b} □
```

## 1.2 Arrays

We assume in the sequel that x, y, z are defined as array[1..n] of integer (here n is a fixed positive integer constant) unless otherwise stated.

**1.2.1.** Fill the array x with zeros. (Write a program fragment whose execution guarantees that all values x[1]..x[n] are zero independent of the initial value of x.)

*Solution.*

```
i := 0;
{invariant relation: x[1],..,x[i] = 0}
while i <> n do begin
  i := i + 1;
  {x[1]..x[i-1] = 0}
  x[i] := 0;
end;
                                                              □
```

**1.2.2.** Count the number of zeros in an array x. (Write a program fragment that does not change the value of x and guarantees that the integer variable k contains the number of zeros among x[1]..x[n].)

*Solution.*

```
    . . .
{invariant: k = number of zeros among x[1]..x[i] }
    . . .
                                                              □
```

**1.2.3.** Not using assignment statement for arrays, write a program that is equivalent to the statement x:=y.

*Solution.*

```
i := 0;
{invariant: y is unchanged, x[t]=y[t] for all t<=i}
while i <> n do begin
  i := i + 1;
```

```
{x[t]=y[t] for all t<i}
x[i] := y[i];
end;                                                                □
```

**1.2.4.** Find the maximum value among x[1]..x[n].

*Solution.*

```
i := 1; max := x[1];
{invariant relation: max = maximum(x[1]..x[i])}
while i <> n do begin
  i := i + 1;
  {max = maximum(x[1]..x[i-1])}
  if x[i] > max then begin
    max := x[i];
  end;
end;                                                                □
```

**1.2.5.** An array x: array[1..n] of integer is given such that x[1] $\leqslant$ x[2] $\leqslant \ldots \leqslant$ x[n]. Find the number of different elements among x[1]..x[n].

*Solution.* (Version 1)

```
i := 1; k := 1;
{invariant relation: k = the number of
 different elements among x[1]..x[i]}
while i <> n do begin
  i := i + 1;
  if x[i] <> x[i-1] then begin
    k := k + 1;
  end;
end;
```

(Version 2) The number in question is the number of i in 1..n−1 such that x[i] is not equal to x[i+1], plus one.

```
k := 1;
for i := 1 to n-1 do begin
  if x[i]<> x[i+1] then begin
    k := k + 1;
  end;
end;                                                                □
```

**1.2.6.** An array x: array[1..n] of integer is given. Compute the number of different elements among x[1]..x[n]. (The number of operations should be of order $n^2$.)                                                                □

**1.2.7.** The same problem with an additional requirement: the number of operations should be of order $n \log n$.

[*Hint.* See chapter 4 on sorting.] □

**1.2.8.** The same problem where all elements are integers in 1..k and the number of operations should be of order n + k. □

**1.2.9.** (Communicated by A.L. Brudno.) A rectangular field m × n contains mn squares. Some squares are marked as black. It is known that black squares are grouped into several disjoint rectangles (no common sides). Assuming that the colors of squares are represented as

```
array [1..m] of array [1..n] of Boolean;
```

count the number of rectangles. The number of operations should be of order mn.

*Solution.* The number of rectangles is equal to the number of their upper left corners. It is easy to check whether a square is in the upper left corner. Just check the color of the cell as well as the colors of its upper and left neighbors. (Don't forget the case when the cell is on the left or upper boundary of a given m × n rectangle.) □

**1.2.10.** An array x[1]..x[n] is given. Without using other arrays, put its elements in reverse order.

*Solution.* We should exchange x[i] and x[n+1-i] for all indices i such that i < n + 1 - i, i.e., $2i < n + 1 \Leftrightarrow 2i \leqslant n \Leftrightarrow i \leqslant n$ div 2:

```
for i := 1 to n div 2 do begin
| ...exchange x[i] and x[n+1-i];
end;
```
□

**1.2.11.** (From D. Gries' book [7]) An array x[1]..x[m+n] is considered as a concatenation of two segments: a prefix x[1]..x[m] of length m and a suffix x[m+1]..x[m+n] of length n. Without using other arrays, exchange these prefix and suffix segments. (The number of operations should be of order m + n.)

*Solution.* (Version 1) Reverse the prefix segment (see the preceding problem), then the suffix segment, and finally the whole array.

(Version 2, A.G. Kushnirenko) Imagine that the array is written down along a circle. Then the required transformation is a rotation. Recall that rotation may be represented as the composition of two axial symmetries. Each symmetry can be performed by exchanges without extra memory.

(Version 3) Consider a more general problem: Exchange two adjacent segments x[p+1]..x[q] and x[q+1]..x[r] in an array. Assume that the length of the left segment (called $A$ in the sequel) does not exceed the length of the right segment (called $B$). Split $B$ into two segments $B_1$ and $B_2$, where $B_1$ is an initial segment of $B$ of the same length as $A$. (So, $B = B_1 + B_2$, where + stands for concatenation.) We need to transform $A + B_1 + B_2$ into $B_1 + B_2 + A$. We can easily exchange $A$ and $B_1$ because they have equal lengths. After that we get $B_1 + A + B_2$ and it remains to exchange $A$ and $B_2$. Therefore, we have reduced our problem to a similar one for shorter segments. Here is the outline of the program:

```
p := 0; q := m; r := m + n;
{invariant relation: it remains
   to exchange x[p+1..q], x[q+1..r]}
while (p <> q) and (q <> r) do begin
   {both segments are nonempty}
   if (q - p) <= (r - q) then begin
      ..exchange x[p+1]..x[q] and x[q+1]..x[q+(q-p)]
      pnew := q; qnew := q + (q - p);
      p := pnew; q := qnew;
   end else begin
      ..exchange x[q-(r-q)+1]..x[q] and x[q+1]..x[r]
      qnew := q - (r - q); rnew := q;
      q := qnew; r := rnew;
   end;
end;
```

The number of operations may be estimated as follows. At each step the part of the array that should be processed becomes shorter by the length of $A$. The number of operations required is also proportional to the length of $A$. □

**1.2.12.** An array a: `array[0..n]` of `integer` contains the coefficients of a polynomial of degree n. Compute the value of this polynomial at the point x; that is, `a[n] x^n + ⋯ + a[1] x + a[0]`.

*Solution.* (The algorithm described below is called Horner's rule.)

```
k := 0; y := a[n];
{invariant relation: 0 <= k <= n,
 y= a[n]*(x ** k)+...+a[n-1]*(x ** (k-1))+...+
                     + a[n-k]*(x ** 0)}
while k <> n do begin
   k := k + 1;
   y := y * x + a [n-k];
end;
```
□

**1.2.13.** (Requires some calculus; communicated by A.G. Kushnirenko) Extend Horner's rule to compute not only the value of a polynomial at some point, but also the value of the derivative of the same polynomial at the same point.

*Solution.* When a new coefficient is added, the polynomial changes from $P(x)$ to $Q(x) = x P(x) + c$. The derivative $Q'(x)$ is equal to $x P'(x) + P(x)$. Therefore we can easily compute $Q(x)$ and $Q'(x)$ if we know $x, c, P(x)$ and $P'(x)$. □

This solution has a unexpected feature: we do not need to know in advance the degree of the polynomial..(An attempt to use it makes the solution only more complicated, especially if we try to compute only the value of the derivative.)

There is a general statement about the computation of derivatives:

**1.2.14.** (W. Baur, V. Strassen) Assume that a "straight-line" program (i.e., a program containing only assignment statements) computes the value of some polynomial $P(x_1, \ldots, x_n)$ given the variables $x_1, \ldots, x_n$. We assume that the right-hand sides of the assignment statements are expressions that contain only addition, multiplication, constants, variables $x_1, \ldots, x_n$ and the variables that appear on the left-hand side of previous assignment statements. Prove that there exists a program of the same type that computes all $n$ derivatives $\partial P / \partial x_1, \ldots, \partial P / \partial x_n$, and the number of arithmetic operations is only $C$ times larger than in the original program. (Here the constant $C$ does not depend on $n$.)

[*Hint.* We may assume that each assignment consists of addition, multiplication by a constant, or multiplication of two variables. Use induction on the number of statements, applying the inductive assumption to the program obtained by deleting the *first* assignment of the program.]  □

**1.2.15.** Two arrays a: `array[0..k] of integer` and b: `array[0..1] of integer` contain the coefficients of two polynomials of degrees k and 1 respectively. Put into c: `array[0..m] of integer` the coefficients of their product. (Here k, 1, m are nonnegative integers such that m = k + 1; the array element indexed by i contains the coefficient of $x^i$.)

*Solution.*

```
for i:=0 to m do begin
| c[i]:=0;
end;
for i:=0 to k do begin
| for j:=0 to 1 do begin
| | c[i+j] := c[i+j] + a[i]*b[j];
| end;
end;
```
□

**1.2.16.** The polynomial multiplication algorithm given above uses about $n^2$ operations to compute the product of two polynomials of degree $n$. Find an (asymptotically) more effective algorithm that uses only $O(n^{\log 4 / \log 3})$ operations.

[*Hint.* Suppose we want to multiply two polynomials of degree $2k$. Represent these polynomials as

$$A(x) x^k + B(x) \quad \text{and} \quad C(x) x^k + D(x)$$

where $A, B, C, D$ are polynomials of degree $k$. The product in question is equal to

$$A(x)C(x) x^{2k} + (A(x)D(x) + B(x)C(x)) x^k + B(x)D(x).$$

The natural way to compute $AC$, $AD + BC$, $BD$ requires four multiplications of degree $k$ polynomials. However, the following trick requires only three multiplications: compute $AC$, $BD$ and $(A + B)(C + D)$, then use the identity $AD + BC = (A + B)(C + D) - AC - BD$.]  □

**1.2.17.** Two arrays x: array[1..k] of integer and y: array[1..1] of integer are sorted (x[1] < ... < x[k], y[1] < ... < y[1]). Find the number of common elements in both arrays; that is, the number of integers t such that t = x[i] = y[j] for some i and j. (The number of operations should be of order k + 1.)

*Solution.*

```
    k1:=0; 11:=0; n:=0;
    {invariant relation: 0<=k1<=k; 0<=11<=1;
     the number in question is n plus the number of common
     elements in x[k1+1],...,x[k] and y[11+1],...,y[1]}
    while (k1 <> k) and (11 <> 1) do begin
      if x[k1+1] < y[11+1] then begin
      | k1 := k1 + 1;
      end else if x[k1+1] > y[11+1] then begin
      | 11 := 11 + 1;
      end else begin {x[k1+1] = y[11+1]}
      | k1 := k1 + 1;
      | 11 := 11 + 1;
      | n := n + 1;
      end;
    end;
    {k1 = k or 11 = 1; therefore, one of the sets
     mentioned in the invariant relation is empty
     and n is the number in question}
```

*Remark.* In the last alternative it is enough to increase only one of the variables k1 and 11 (though the symmetry would be broken if we did that).            □

**1.2.18.** Solve the preceding problem with the assumption that x[1] ⩽ ... ⩽ x[k] and y[1] ⩽ ... ⩽ y[1] (arrays are non-decreasing but not necessarily increasing).

*Solution.* In the third alternative of the previous solution, when increasing k1 and 11 by 1, we decreased (by 1) the number of common elements in x[k1+1] ... x[k] and x[11+1] ... x[1]. For non-decreasing arrays, this is not enough since the same element may appear many times. A more complicated procedure is required:

```
    ...
    end else begin {x[k1+1] = y[11+1]}
    | t := x [k1+1];
    | while (k1<k) and (x[k1+1]=t) do begin
    | | k1 := k1 + 1;
    | end;
    | while (11<1) and (x[11+1]=t) do begin
    | | 11 := 11 + 1;
    | end;
```

```
  | n := n+1;
  end;
```

*Remark.* This program has a bug, however. If in the condition

$$(k1<k) \text{ and } (x[k1+1]=t)$$

(or in the similar second condition) the first expression (k1<k) is false, the second one is meaningless (index out of bounds) and an error may occur. Some versions of Pascal use "short circuit evaluation" of Boolean expressions: when evaluating A and B the evaluation of B is "short circuited" when A is false. In this case, the problem disappears.

Rather than rely on implementation-dependent features (short-circuit evaluation is not prescribed by the Pascal's author, N. Wirth), we can do the following. Introduce an additional variable b: Boolean and write:

```
if k1 < k  then b := (x[k1+1]=t)  else  b:=false;
{b = (k1<k) and (x[k1+1] = t)}
while  b  do  begin
  | k1 := k1+1;
  | if k1 < k then b := (x[k1+1]=t)  else  b:=false;
end;
```

Another possibility (which is shorter, but less symmetric):

```
end else begin {x[k1+1] = y[l1+1]}
  | if k1 + 1 = k then begin
  |   | k1 := k1 + 1;
  |   | n := n + 1;
  | end else if x[k1+1] = x [k1+2] then begin
  |   | k1 := k1 + 1;
  | end else begin
  |   | k1 := k1 + 1;
  |   | n := n + 1;
  | end;
end;
```

Alternatively, we can increase the constant in the array declaration and reserve a spare memory location. □

**1.2.19.** Two arrays x: array[1..k] of integer and y: array[1..l] of integer satisfying $x[1] \leqslant \ldots \leqslant x[k], y[1] \leqslant \ldots \leqslant y[l]$ are given. Find the number of different elements among $x[1], \ldots, x[k], y[1], \ldots, y[l]$. (The number of operations should be of order $k + l$.) □

**1.2.20.** Two arrays $x[1] \leqslant \ldots \leqslant x[k]$ and $y[1] \leqslant \ldots \leqslant y[l]$ are given. Merge them into one array $z[1] \leqslant \ldots \leqslant z[m]$ ($m = k + l$). Any element should appear is z as many times as it appears in x and y together. The number of operations should be of order m.

*Solution.*

```
k1 := 0; l1 := 0;
{invariant relation: the answer is the concatenation
  of z[1]..z[k1+l1] and the merge of
  x[k1+1]..x[k] and y[l1+1]..y[l]}
while (k1 <> k) or (l1 <> l) do begin
  if k1 = k then begin
    {l1 < l}
    l1 := l1 + 1;
    z[k1+l1] := y[l1];
  end else if l1 = l then begin
    {k1 < k}
    k1 := k1 + 1;
    z[k1+l1] := x[k1];
  end else if x[k1+1] <= y[l1+1] then begin
    k1 := k1 + 1;
    z[k1+l1] := x[k1];
  end else if x[k1+1] >= y[l1+1] then begin
    l1 := l1 + 1;
    z[k1+l1] := y[l1];
  end else begin
    {this cannot happen}
  end;
end;
{k1 = k, l1 = l, arrays are merged}
```

This process can be illustrated as follows. Assume we have two piles of cards with a word on each card, and each pile is alphabetically sorted. We merge them into one pile as follows. At every step we compare the first cards of both piles and take the one which is alphabetically first. If one pile is already empty, we take the remaining cards from the other pile.

**1.2.21.** Two arrays $x[1] \leqslant \ldots \leqslant x[k]$ and $y[1] \leqslant \ldots \leqslant y[l]$ are given. Find their "intersection", i.e., an array $z[1] \leqslant \ldots \leqslant z[m]$ that contains their common elements. The multiplicity of each element in z should be equal to the smaller of its multiplicities in x and y. The number of operations should be of order $k + l$. □

**1.2.22.** Two arrays $x[1] \leqslant \ldots \leqslant x[k]$ and $y[1] \leqslant \ldots \leqslant y[l]$ and a number q are given. Find i and j such that $x[i] + y[j]$ is as close to q as possible. (The number of operations should be of order k+l. You may use a fixed number of auxiliary integer variables; the arrays x and y are read-only.)

[*Hint.* We need to find the minimal distance between $x[1] \leqslant \ldots \leqslant x[k]$ and $q - y[l] \leqslant \ldots \leqslant q - y[1]$. This is easily done while merging these numbers into one (imaginary) array.] □

**1.2.23.** (from D. Gries' book [7]) There is a number that is present in all three non-decreasing arrays x[1] $\leqslant \ldots \leqslant$ x[p], y[1] $\leqslant \ldots \leqslant$ y[q], z[1] $\leqslant \ldots \leqslant$ z[r]. Find this number (or one of them, if there is more than one). The number of operations should be of order p + q + r.

*Solution.*

```
p1:=1; q1=1; r1:=1;
{invariant relation: x[p1]..x[p], y[q1]..y[q],
 z[r1]..z[r] have an element in common}
while not ((x[p1]=y[q1]) and (y[q1]=z[r1])) do begin
  if x[p1]<y[q1] then begin
  | p1:=p1+1;
  end else if y[q1]<z[r1] then begin
  | q1:=q1+1;
  end else if z[r1]<x[p1] then begin
  | r1:=r1+1;
  end else begin
  | {this cannot happen}
  end;
end;
{x[p1] = y[q1] = z[r1]}
writeln (x[p1]);                                              □
```

**1.2.24.** Repeat the previous problem assuming that we do not know in advance if such a common element exist. Determine whether or not it exists and locate it if it does.                                                                    □

**1.2.25.** The array a[1..n] consists of arrays [1..m] of integers:

$$a: array [1..n] of array [1..m] of integer;$$

$$a[1][1] \leqslant \ldots \leqslant a[1][m], \ldots, a[n][1] \leqslant \ldots \leqslant a[n][m].$$

It is known that there is a common number present in all a[i] (that is, there exists an x such that for all i in 1..n there exists a j in 1..m such that a[i][j] = x). Find such a number x.

*Solution.* We use an array b[1]..b[n] whose elements mark the start of the "non-scanned" portions of arrays a[1], ..., a[n].

```
for k:=1 to n do begin
| b[k]:=1;
end;
eq := true;
for k := 2 to n do begin
| eq := eq and (a[1][b[1]] = a[k][b[k]]);
end;
```

```
{invariant relation: non-scanned parts have nonempty
 intersection, i.e., there is x such that for any i in
 [1..n] there is j in [b[i]..m] such that a[i][j] = E;
 eq <=> first non-scanned elements are all equal}
while not eq do begin
  s := 1; k := 1;
  {a[s][b[s]] is minimal among a[1][b[1]]..a[k][b[k]]}
  while k <> n do begin
    k := k + 1;
    if a[k][b[k]] < a[s][b[s]] then begin
      s := k;
    end;
  end;
  {a[s][b[s]] is minimal among a[1][b[1]]..a[n][b[n]]}
  b [s] := b [s] + 1;
  for k := 2 to n do begin
    eq := eq and (a[1][b[1]] = a[k][b[k]]);
  end;
end;
writeln (a[1][b[1]]);                                    □
```

**1.2.26.** Our solution of the preceding problem requires $mn^2$ operations. Find an algorithm that needs only $O(mn)$ operations (i.e., not more than $Cmn$ operations for some $C$).

[*Hint.* We have to break the symmetry and choose one of the rows as a "principal" row. We move along the principal row maintaining the following relation: in all other rows the maximal element not exceeding the current element of the principal row is located.]                                                                    □

**1.2.27.** (Binary search) An array $x[1] \leqslant \ldots \leqslant x[n]$ of integers and an integer a are given. Determine if a is present in x; that is, if there exists an i in 1..n such that $x[i] = a$. (The number of operations should be of order $\log n$.)

*Solution.* (We assume that $n > 0$.)

```
l := 1; r := n+1;
{r > l, if a appears, it appears among x[1]..x[r-1]}
while r - l <> 1 do begin
  m := l + (r-l) div 2 ;
  {l < m < r }
  if x[m] <= a then begin
    l := m;
  end else begin {x[m] > a}
    r := m;
  end;
end;
```

(Check that the invariant relation is maintained even if x[m] = a.)

At each step the difference $r - 1$ is halved, so we get the required bound for the number of operations.

Program can be simplified using the equality

$$1 + (r-1) \operatorname{div} 2 = (21 + (r - 1)) \operatorname{div} 2 = (r + 1) \operatorname{div} 2. \qquad \square$$

*Remark.* It is very important that the array x[1]..x[n] is sorted; otherwise we obviously have to test all n elements x[1]..x[n] to be sure that a given element is not in the array ("sequential search").

**1.2.28.** (From D. Gries' book [7]) An array

```
x: array[1..n] of array[1..m] of integer
```

is sorted both row-wise and column-wise:

$$x[i][j] \leqslant x[i][j+1],$$
$$x[i][j] \leqslant x[i+1][j].$$

Determine if a given number a is present among the array elements x[i][j].

*Solution.* Represent x as a rectangular matrix. Choose a rectangle that contains a (assuming that a is present at all) and then make this rectangle smaller and smaller. This rectangle contains x[i][j] such that $1 \leqslant i \leqslant 1$ and $k \leqslant j \leqslant m$.

(The rectangle is empty if $1 = 0$ or $k = m + 1$.)

```
l:=n; k:=1;
{l>=0, k<=m+1, if a appears at all, it appears
  inside the rectangle}
while (l>0) and (k<m+1) and (x[l][k]<>a) do begin
  if x[l][k] < a then begin
  | k := k + 1; {left column cannot contain a, delete it}
  end else begin {x[l][k] > a}
  | l := l - 1; {last row cannot contain a, delete it}
  end;
end;
{x[l][k] = a or the rectangle is empty}
answer:= (l > 0) and (k < m+1) ;
```

*Remark.* Here the same error as in problem 1.2.18 appears: x[l][k] may be undefined. (We leave its correction to the reader.)                                              □

**1.2.29.** (Moscow programming contest) A non-decreasing integer array a[1] ⩽ a[2] ⩽ ... ⩽ a[n] contains positive numbers only. Find the minimal positive integer that cannot be represented as a sum of several elements of this array (no element may be used more than once). The number of operations should be of order n.

*Solution.* Assume all numbers that can be represented as sums of subsets of {a[1],...,a[k]} form the set {1, 2, ..., N} for some N. If a[k+1] is greater than N+1, then N+1 is the smallest number that cannot be represented as the sum of some subset of {a[1],...,a[n]}. If a[k+1] ⩽ N+1, then all numbers that can be represented as sums of subsets of {a[1] ... a[k+1]} form the set {1, 2, ..., N+a[k+1]}.

```
k := 0; N := 0;
{invariant relation: all the numbers that can be
 represented as sums of subsets of {a[1],..,a[k]},
 form the set {1,2,...,N}}
while (k <> n) and (a[k+1] <= N+1) do begin
| N := N + a[k+1];
| k := k + 1;
end;
{(k = n) or (a[k+1] > N+1); the answer is N+1
 in both cases}
writeln (N+1);
```

(Error: when the first condition in the while-construct is false, the second is undefined.)                                                                                          □

**1.2.30.** (Requires some algebra) An integer array a[1]..a[n] contains some permutation of 1..n (each of numbers 1..n appears exactly once).

(a) Determine if the permutation is even.

(b) Without using other arrays, replace the permutation by its inverse permutation (i.e., if a[i] = j was true before execution, then a[j] = i is true after execution).

(In both (a) and (b), the number of operations should be of order n.)

[*Hint.* (a) The number of cycles determines whether a permutation is even or odd. To mark an already counted cycle, we can (for example) change the sign of its elements. (b) The inverse permutation is computed cycle by cycle.]                          □

**1.2.31.** An array a[1..n] and a threshold b are given. Rearrange the elements of the array in such a way that all elements on the left of some boundary do not exceed b whereas all elements on the right of the boundary are greater than or equal to b. The number of operations should be proportional to n.

*Solution.*

```
l:=0; r:=n;
{invariant relation: l<=r; a[1..l]<=b; a[r+1..n]>=b}
while l <> r do begin
  if a[l+1] <= b then begin
  | l:=l+1;
  end else if a[r] >=b then begin
  | r:=r-1;
  end else begin {a[l+1]>b; a[r]<b; l+1<r}
  | ..exchange a[l+1] and a[r]
  | l:=l+1; r:=r-1;
  end;
end;
```
□

**1.2.32.** Repeat the previous problem with the additional restriction that the elements smaller than b should precede elements equal to b which themselves should precede elements greater than b.

*Solution.* We need three boundaries to divide our segment into four parts. The first part contains elements smaller than b; the second part contains only elements equal to b; the third part may contain anything; and the fourth part contains only elements greater than b. (We can get a more symmetric solution using a fourth boundary.) At each step we consider the left element of the third part (just to the right of the second boundary).

```
l:=0; m:=0; r:=n;
{invariant relation: a[1..l]<b; a[l+1..m]=b; a[r+1..n]>b}
while m <> r do begin
  if a[m+1]=b then begin
  | m:=m+1;
  end else if a[m+1]>b then begin
  | ..exchange a[m+1] and a[r]
  | r:=r-1;
  end else begin {a[m+1]<b}
  | ..exchange a[m+1] and a[l+1]
  | l:=l+1; m:=m+1;
  end;
end;
```
□

**1.2.33.** (This version of the preceding problem is called the "Dutch flag" problem in E. Dijkstra's book [5].) The array contains n elements; each element is equal to 0, 1, or 2. Sort the array if the only allowed operation (besides reading its elements) is the exchange of two elements of the array. The number of operations should be proportional to n. □

**1.2.34.** An array a[1..n] and a number $m \leqslant n$ are given. For any segment formed by m adjacent elements (there are $n - m + 1$ segments of this type) compute its sum. The total number of operations should be of order n.

*Solution.* When moving the segment to the right, add one element and subtract another.                                                                                    □

**1.2.35.** A square matrix a[1..n][1..n] and a number m ≤ n are given. For any m×m-subsquare, compute the sum of its elements. The total number of operations should be of order $n^2$.

*Solution.* First compute the sum for all horizontal rectangles of size m × 1. (When such a rectangle is shifted to the right, one element is added and one is subtracted.) After computing all these sums, we compute the sums for squares. (When a square is shifted down, one rectangle is added and another rectangle is subtracted.)         □

**1.2.36.** An array a[1]..a[n] contains all integers in [0..n] except one. Find this omitted integer with fixed additional memory. Number of operations should be proportional to n.

[*Hint.* Compute the sum of all elements.]                                                □

**1.2.37.** An array a[1]..a[n] contains some integers, and every element appears twice except for one element that appears only once. Find this element with fixed additional memory. Number of operations should be proportional to n.

[*Hint.* Use xor operation.]                                                              □

## 1.3 Inductive functions (following A.G. Kushnirenko)

Let M be a set. Let f be a function whose arguments are finite sequences of elements of M and whose values belong to some other set N. The function f is called *inductive* if its value on the sequence x[1]..x[n] is uniquely determined by its value on the sequence x[1]..x[n-1] and by x[n]; that is, if there is a function F : N × M → N such that

$$f(\langle x[1],\ldots,x[n]\rangle) = F(f(\langle x[1],\ldots,x[n-1]\rangle),x[n]).$$

For example, the sum x[1]+···+x[n] is an inductive function (it is enough to know the sum x[1] + ··· + x[n-1] and the value of x[n] to compute x[1] + ··· + x[n]). At the same time, the average value is not an inductive function; if we know x[n] and the average of x[1], ..., x[n-1], but have no information about n, we cannot compute the average of x[1], ..., x[n].

An inductive function can be computed as follows:

```
k := 0; f := f0;
{invariant relation:
   f is a value of the function on <x[1],...,x[k]>}
while  k<>n do begin
   k := k + 1;
   f := F (f, x[k]);
end;
```

Here f0 is the value of the function on the empty sequence (sequence of length 0). If f is defined only on nonempty sequences, the first line should be replaced by

$$k:=1; \quad f:=f(<x[1]>);$$

If a given function f is not inductive, it is instructive to look for its *inductive extension*. By an inductive extension of f we mean an inductive function g whose values determine uniquely the values of f, i.e., there exists a function t such that

$$f(\langle x[1] \dots x[n] \rangle) = t(g(\langle x[1] \dots x[n] \rangle))$$

for all $\langle x[1] \dots x[n] \rangle$. One can prove that there exists a minimal extension F among all inductive extensions of a given function f. Here the word "minimal" means that for any other inductive extension g the values of F are determined uniquely by the values of g; that is, $F(x) = u(g(x))$ for some function u.

**1.3.1.** Find an inductive extension for the following functions:

(a) the average value of a sequence of real numbers;

(b) the number of elements in a sequence that are equal to its maximal element;

(c) the second largest element of the sequence (second from the top after the sequence is sorted in non-descending order);

(d) the maximal number of consecutive equal elements;

(e) the maximal length of a monotone (non-increasing or non-decreasing) fragment composed of consecutive elements of a sequence;

(f) the number of groups of ones separated by zeros (in a 0-1-sequence).

*Solution.*

(a) As we have seen, the average value is not an inductive function. However, the average value is a ratio of two inductive functions. The first one is the sum of all the terms; the second one is the number of terms. Therefore, the combination ⟨the sum of all elements; the length⟩ is an inductive extension.

(b) ⟨the maximal element; the number of elements equal to the maximal element⟩;

(c) ⟨the maximal element; the second maximal element⟩;

(d) ⟨the maximal number of adjacent equal elements; the maximal number of adjacent equal elements at the end of the sequence; the last element⟩;

(e) ⟨the maximal length of a monotone fragment; the maximal length of a non-decreasing fragment at the end of the sequence; the maximal length of a non-increasing fragment at the end of the sequence; the last term of the sequence⟩;

(f) ⟨the number of 1-groups; the last term⟩.  □

**1.3.2.** (Communicated by D.V. Varsonofiev) Two sequences x[1]..x[n] and y[1]..y[k] of integers are given. Determine if the second sequence is a subsequence of the first one; that is, if it is possible to delete some terms of the first sequence to obtain the second one. The number of operations should be of order n + k.

*Solution.* (Version 1) Reduce the problem to the same problem involving shorter sequences.

```
n1:=n;
k1:=k;
{invariant relation: the answer is TRUE <=>
   it is possible to get y[1]..y[k1] out of x[1]..x[n1]}
while (n1 > 0) and (k1 > 0) do begin
   if x[n1] = y[k1] then begin
      n1 := n1 - 1;
      k1 := k1 - 1;
   end else begin
      n1 := n1 - 1;
   end;
end;
{n1 = 0 or k1 = 0; if k1 = 0, the answer is positive;
 if k1<>0 (and n1 = 0), the answer is negative}
answer := (k1 = 0);
```

We use the following fact: If $x[n1] = y[k1]$ and the sequence $y[1]..y[k1]$ is a subsequence of $x[1]..x[n1]$, then $y[1]..y[k1-1]$ is a subsequence of $x[1]..x[n1-1]$.

(Version 2) The function $\langle x[1]..x[n1]\rangle \mapsto$ (the maximal value of k1 such that $y[1]..y[k1]$ is a subsequence of $x[1]..x[n1]$) is inductive. □

**1.3.3.** Two sequences $x[1]..x[n]$ and $y[1]..y[k]$ of integers are given. Find the maximal length of a sequence that is a subsequence of both given sequences. The number of operations should be of order $n \cdot k$.

*Solution* (communicated by M.N. Weinzweig and A.M. Dimentman). Denote the maximal length of a common subsequence of sequences $x[1]..x[p]$ and $y[1]..y[q]$ by $f(p, q)$. Then

$$x[p] \neq y[q] \Rightarrow f(p, q) = \max(f(p, q-1), f(p-1, q));$$
$$x[p] = y[q] \Rightarrow f(p, q) = \max(f(p, q-1), f(p-1, q), f(p-1, q-1)+1)$$

In the second case, the maximum of three numbers is in fact equal to the third number because $f(p-1, q-1) + 1 \geqslant f(p, q-1), f(p-1, q)$.

Therefore we can construct a table of f-values. This table is of size $n \cdot k$. We can even proceed using only k (or n) memory locations if we compute (for $p = 1, 2, \ldots$) the array $\langle f(p, 0), \ldots, f(p, k)\rangle$ (it is an inductive function of p). □

**1.3.4.** (From D. Gries' book [7]) A sequence of integers $x[1], \ldots, x[n]$ is given. Find the maximum length of its increasing subsequence. (The number of operations should be of order $n \log n$).

*Solution.* The function in question is not inductive. However, it has the following inductive extension: it consists of the maximal length of an increasing subsequence (denoted by k in the sequel) and the numbers $u[1], \ldots, u[k]$, where $u[i]$ is the minimal last term of all increasing subsequences of length i. Evidently, $u[1] \leqslant$

... $\leqslant$ u[k]. When a new term is appended to x, the values of u and k should be updated.

```
n1 := 1; k := 1; u[1] := x[1];
{invariant: k and u satisfy the description above}
while n1 <> n do begin
  n1 := n1 + 1;
  ...
  {i is the maximal number in 1..k such that
    u[i] < x[n1]; i=0 if there is no such numbers}
  if i = k then begin
    k := k + 1;
    u[k+1] := x[n1];
  end else begin {i < k, u[i] < x[n1] <= u[i+1] }
    u[i+1] := x[n1];
  end;
end;
```

The omitted fragment employs binary search (see 1.2.27, p. 24). In the invariant relation we assume that u[0] $= -\infty$ and u[k+1] $= +\infty$. The goal is u[i] $<$ x[n1] $\leqslant$ u[i+1].

```
i:=0; j:=k+1;
{u[i] < x[n1] <= u[j], j > i}
while (j - i) <> 1 do begin
  s := i + (j-i) div 2;    {i < s < j}
  if x[n1] <= u[s] then begin
    j := s;
  end else begin {u[s] < x[n1]}
    i := s;
  end;
end;
{u[i] < x[n1] <= u[j], j-i = 1}
```

*Remark.* We get a simpler (but not minimal) inductive extension if, for any i, we keep the maximal length of an increasing sequence whose last term is x[i]. This extension leads to an algorithm requiring $n^2$ operations.

Another nice quadratic algorithm (communicated by M. Vyugin): look for the longest common subsequence in the original array x and the sorted array using the preceding problem.  □

**1.3.5.** What changes are needed in the solution of the previous problem if we are looking for a maximal *non-decreasing* sequence?  □

# 2

## Generation of combinatorial objects

In this chapter, we deal with problems that require us to generate all the elements of some finite set one-by-one. We start with a simple example in section 2.1 (generating all sequences of fixed length composed of elements of some finite set). Then in section 2.2 we generate all permutations of a given set. It is more difficult since now the elements are not independent (no element should appear twice). Two other popular combinatorial objects are considered in sections 2.3 (subsets of fixed size) and 2.4 (partitions). Some applications (including Gray codes) are considered in section 2.5. In section 2.6 we consider some examples where elements to be generated are in one-to-one correspondence with elements of some other set (which are easier to generate). Finally, in section 2.7 we consider a classical problem where we have to count elements of some class (without generating them).

## 2.1 Sequences

**2.1.1.** Print all the sequences of length k composed of the numbers 1..n.

*Solution.* Let us print them in lexicographic order (a sequence a precedes sequence b if for some s their initial segments of length s are equal and the (s+1)-th term of a is smaller.) The first sequence in this ordering is <1,1,...,1>; the last one is <n,n,...,n>. We use an array x[1]..x[k] to store the last sequence printed.

```
..make x[1]...x[k] equal to 1
..print x
..make last[1]...last[k] equal to n
{all sequences up to x (including x) are printed}
while x <> last do begin
  ...x := the successor of x
  ...print x
end;
```

A. Shen, *Algorithms and Programming*, Springer Undergraduate Texts in Mathematics and Technology, DOI 10.1007/978-1-4419-1748-5_2,

Let us explain how to get the successor of x. By definition, the successor should have the same first s terms and larger (s+1)-th term. This is possible only if x[s+1] < n. To get the immediate successor, we find the maximal s with this property and increase the corresponding element by 1. In other words, we move along the sequence from right to left and find the rightmost term that is smaller than n (it does exist, because x<>last by assumption). Then we increase it by 1 and make all the subsequent terms equal to 1.

```
p := k;
while not (x[p] < n) do begin
| p := p-1;
end;
{x[p] < n, x[p+1] =...= x[k] = n}
x[p] := x[p] + 1;
for i := p+1 to k do begin
| x[i] := 1;
end;
```

*Remark.* If we use the numbers 0..n-1 instead of 1..n, then finding the successor corresponds to adding 1 in n-ary notation.                     □

**2.1.2.** The program above uses comparisons for arrays (x <> last). Eliminate this step by using a Boolean variable is_last and adding the requirement

$$\text{is\_last} \Leftrightarrow x = \text{last}$$

to the invariant relation.                                                □

**2.1.3.** Print all subsets of the set {1..k}.

*Solution.* These subsets are in one-to-one correspondence with all sequences of 0s and 1s of length k.                                               □

**2.1.4.** Print all sequences of length k of positive integers such that the i-th term does not exceed i for all i.                                    □

## 2.2 Permutations

**2.2.1.** Print all permutations of 1..n (i.e., all sequences of length n that contain all the numbers 1...n).

*Solution.* We store the current permutation in an array x[1]..x[n]. Permutations are printed in lexicographic order. The first permutation (in this order) is ⟨1 2..n⟩. The last one is ⟨n..2 1⟩. How do we find the next permutation in the lexicographic order? When is it possible to increase k-th term in a permutation without changing all preceding terms? The answer is: When the term is smaller than one of the subsequent terms (i.e., terms with larger indices). Therefore, to find the next permutation we should find the maximum k where increase is possible; that is, a k such that

$$x[k] < x[k+1] > \ldots > x[n]$$

Next we increase x[k] but keep the increase as small as possible. This means that we must find the minimal number among x[k+1]..x[n] that is larger than x[k]. After we exchange x[k] with the number found, we have to rearrange x[k+1]..x[n] to make the permutation as small as possible. To achieve this goal, we put x[k+1]..x[n] in increasing order. (Fortunately, they are already arranged in *decreasing* order.)

Here's how to get the next permutation:

```
{<x[1]...x[n]> <> <n...2,1>}
k:=n-1;
{after x[k] terms go in decreasing order: x[k+1]>...>x[n]}
while x[k] > x[k+1] do begin
| k:=k-1;
end;
{x[k] < x[k+1] > ... >  x[n]}
t:=k+1;
{t<=n, all terms x[k+1] > ... > x[t] are bigger than x[k]}
while (t < n) and (x[t+1] > x[k]) do begin
| t:=t+1;
end;
{x[k+1] > ... > x[t] > x[k] > x[t+1] > ... > x[n]}
..exchange x[k] and x[t]
{x[k+1] > ... > x[n]}
..put x[k+1]...x[n] in reversed order
```

*Remark.* This program suffers from the usual problem: x[t+1] is undefined when t = n.      □

## 2.3 Subsets

**2.3.1.** Generate all subsets of the set {1..n} having k elements.

*Solution.* Each subset may be represented by a bit string of length n that contains exactly k 1s. (We'll consider another representation later.) We generate these bit strings in lexicographic order. There is a natural way to do this: Generate *all* bit strings and select those that contain exactly k 1s. However, this solution is considered inefficient, because bit strings with k 1s form a tiny fraction of all bit strings of length n. In the program below the generation of each subsequent string requires not more than $C \cdot n$ operations (for some constant $C$).

When is it possible to increase the s-th term of a bit string with k 1s without changing the preceding terms? If x[s] is changed from 0 to 1, we should replace 1 by 0 somewhere to keep the total number of 1s fixed. Therefore, it is necessary to have 1s on the right of x[s].

Conclusion: If we want to find the next bit string with k 1s, we need x[s] to be the rightmost 0 that has some 1s on the right. In this case we have x[s+1]=1 (otherwise, x[s] is not the rightmost one). Therefore, we should look for the maximal s such that x[s]=0 and x[s+1]=1:

$$x \quad \boxed{\phantom{xxxxxxxx} \mid 0 \mid 1..1 \mid 0..0 \mid}$$
$$\uparrow$$
$$s$$

The term x[s+1] may be followed by several 1s and then several 0s. After we replace x[s] by 1 the next terms should be chosen to get the lexicographically first string; that is, 0s should precede 1s. Here is what we get:

first string: $\underbrace{0 \ldots 0}_{n-k}\underbrace{1 \ldots 1}_{k}$

last string: $\underbrace{1 \ldots 1}_{k}\underbrace{0 \ldots 0}_{n-k}$

How to find the next string after x[1]..x[n] (assuming it exists):

```
s := n - 1;
while not ((x[s]=0) and (x[s+1]=1)) do begin
| s := s - 1;
end;
{x[s] should be changed from 0 to 1}
num:=0;
for k := s to n do begin
| num := num + x[k];
end;
{num is the number of 1s among x[s]...x[n], the number
  of 0s is (length - number of 1s); that is, (n-s+1)-num}
x[s]:=1;
for k := s+1 to n-num+1 do begin
| x[k] := 0;
end;
{it remains to put num-1 1s at the end}
for k := n-num+2 to n do begin
| x[k]:=1;
end;                                                                       □
```

We can also represent a subset by a list of its elements. To obtain the unique representation we require that elements should be listed in increasing order. Now we come to the following problem:

**2.3.2.** Generate (in lexicographic order) all increasing sequences of length k consisting of the numbers 1..n. (Example: for n=5, k=2 we get ⟨12 13 14 15 23 24 25 34 35 45⟩.)

*Solution.* The first sequence is ⟨1 2..k⟩; the last one is ⟨(n-k+1)..(n-1) n⟩. When is it possible to increase the s-th element of the sequence? Answer: If it is less than n-k+s. After the s-th element is increased, all subsequent elements should form an arithmetic sequence with difference 1. Here is the algorithm:

```
s:=n;
while not (x[s] < n-k+s) do begin
| s:=s-1;
end;
{s-th element should be increased};
x[s] := x[s]+1;
for i := s+1 to n do begin
| x[i] := x[i-1]+1;
end;                                               □
```

**2.3.3.** Suppose we represent subsets of 1..n of cardinality k by *decreasing* sequences of length k. (Example: ⟨21 31 32 41 42 43 51 52 53 54⟩.) How do we generate these sequences in alphabetical order?

[*Hint.* Find the maximal s such that x[s+1]+1 < x[s]. (If it does not exist, let s=0.) Now increase x[s+1] by 1 and let all subsequent elements be as small as possible (x[t]=k+1-t for t>s).]                                               □

**2.3.4.** Solve the two preceding problems if alphabetic order is replaced by reversed alphabetic order.                                               □

**2.3.5.** Generate all injective mappings of the set {1..k} into {1..n} (assume that k ⩽ n). A mapping is injective if no two elements of 1..k are mapped to the same element of 1..n. Generation of each mapping should require no more that $C \cdot k$ operations.

[*Hint.* This problem can be reduced to generation of permutations and generation of subsets.]                                               □

## 2.4 Partitions

**2.4.1.** Generate all partitions of a given positive integer n; that is, all the representations of n as a sum of positive integers. We do not take the order of the summands into account. (Example: For n=4, partitions are 1+1+1+1, 2+1+1, 2+2, 3+1 and 4.)

*Solution.* Let us agree that (i) the summands are written in non-increasing order; and (ii) the partitions are generated in alphabetic order. We store a partition in the initial part of an array x[1]..x[n]; the length of the partition is k, and the summands are x[1]..x[k]. At the beginning, k = n and all x[1]...x[n] are equal to 1. At the end, x[1] = n and k = 1.

When can we increase x[s] leaving all preceding elements unchanged? This is possible only if x[s-1] > x[s] or if s = 1. Moreover, x[s] may not be the last element of the partition (because an increase in x[s] should be compensated by a decrease in the subsequent elements). After x[s] is increased, all subsequent elements should be chosen as small as possible.

```
s := k - 1;
while not ((s=1) or (x[s-1] > x[s])) do begin
| s := s-1;
end;
{x[s] should be increased}
x[s] := x[s] + 1;
sum := 0;
for i := s+1 to k do begin
| sum := sum + x[i];
end;
{sum = the sum of terms after x[s]}
for i := 1 to sum-1 do begin
| x [s+i] := 1;
end;
k := s+sum-1;                                                    □
```

**2.4.2.** In this problem we also represent partitions as non-increasing sequences, but now we want to generate them in reversed alphabetic order (e.g., for n=4, we would generate 4, 3+1, 2+2, 2+1+1, 1+1+1+1).

[*Hint*. The rightmost term that may be decreased is the rightmost term not equal to 1. Find it and decrease it by 1. All subsequent terms should be taken as large as possible (equal to the selected term when possible; the last one may be smaller).] □

**2.4.3.** Partitions are represented as non-decreasing sequences; generate them in alphabetic order. For example, when n = 4, we would generate 1+1+1+1, 1+1+2, 1+3, 2+2, 4.

[*Hint*. The last term x[k] cannot be increased, but the term x[k-1] can. (Of course, the last one should be decreased to maintain the sum.) If the sequence is no longer non-decreasing, we combine two terms into one. If the sequence is still non-decreasing, then x[k] should be split into several terms equal to x[k-1] (except for the last one, which may be larger).] □

**2.4.4.** Partitions are represented as non-decreasing sequences. Generate them in reversed alphabetic order. (For n = 4 we have 4, 2+2, 1+3, 1+1+2, 1+1+1+1.)

[*Hint.* The element x[s] can be decreased only if s=1 or x[s-1]<x[s]. If x[s] is not the last term, these conditions are sufficient. If it is the last one, then we must also have x[s-1] $\leqslant$ $\lfloor$x[s]/2$\rfloor$ or s=1. (Here $\lfloor\alpha\rfloor$ stands for the integer part of $\alpha$; that is, the greatest integer not exceeding $\alpha$.)]  □

## 2.5 Gray codes and similar problems

Sometimes it is useful to generate objects in an order such that the next object is only a small modification of the preceding one. In this section, we consider several problems of this type.

Consider $2^n$ strings of length $n$ containing only 0s and 1's.

**2.5.1.** Prove that it is possible to list all of them in an order such that two neighboring strings differ only in one bit.

*Solution.* Use induction on $n$. Assume that $x_1, \ldots, x_k$ is such a sequence of $n$-bit strings (here $k = 2^n$; for any $i$, strings $x_i$ and $x_{i+1}$ differ only in one bit). Then the following sequence includes all $(n+1)$-bit strings and satisfies the desired condition:

$$0x_1, 0x_2, \ldots, 0x_k, 1x_k, 1x_{k-1}, \ldots, 1x_1$$

In geometric terms, the problem states that we can traverse the $n$-dimensional Boolean cube visiting each vertex exactly once. The solutions considers $n$-dimensional Boolean cube as composed of two $n - 1$-dimensional Boolean cubes (one may think of top and bottom "faces"); we traverse one of them (using the inductive assumption) and then switch to another one.  □

We'll return to this problem later.

**2.5.2.** Generate all sequences of length n composed of the numbers 1..k in such an order that neighboring sequences differ only in one place, and the numbers at this place differ by 1.

*Solution.* Consider a rectangular chess board of width n and height k. Place a piece in each column of the chess board. The position is represented by a sequence of n integers (each between 1 and k); the i-th number represents a position of the piece in the i-th column. At each piece we draw a small arrow that points up or down. Initially, all the pieces are in the first row and all the arrows point up. We move pieces according to the following rule: Find the rightmost piece that can be moved in the direction of the arrow on it, and move it. At the same time all the pieces on the right (they cannot move in the direction of their arrows) are turned over.

It is evident that at each step only one piece is moving, therefore only one term in the corresponding sequence is changed by 1. Let us prove by induction on n that all sequences of length n composed of the numbers 1..k will appear. The case n=1 is evident, so assume that n > 1. Divide all moves into two categories. The first category is formed by moves where the last (rightmost) piece is moving. The second category is formed by moves where the moving piece is not the last one. In this case the rightmost piece is near the border and is turned over. Therefore, each move of the second category is followed by k-1 moves of the first category; during this period the rightmost piece visits all the cells. Let us forget now about the rightmost piece. Then the first n-1 pieces are moving according to the prescribed rules. Therefore, by the induction assumption, all sequences of length n-1 appear exactly once. The movements of the last piece make k sequences of length n out of each sequence of length n-1.

The program keeps an array x[1]..x[n] (positions of pieces) and an array d[1]..d[n] composed of numbers +1 and -1 (+1 denotes up-arrow; -1 denotes down-arrow).

Initial state: $x[1] = \ldots = x[n] = 1$; $d[1] = \ldots = d[n] = 1$.

The following algorithm produces the next position according to the description above. At the same time, it checks whether the next position exists; the answer is stored in a Boolean variable p.

```
{if possible, make a move and let p := true;
 otherwise, p := false }
 i := n;
 while (i > 1) and
   (((d[i]=1) and (x[i]=n)) or ((d[i]=-1) and (x[i]=1)))
     do begin
     i:=i-1;
 end;
 if (d[i]=1 and x[i]=n) or (d[i]=-1 and x[i]=1)
       then begin
   p:=false;
 end else begin
   p:=true;
   x[i] := x[i] + d[i];
   for j := i+1 to n do begin
     d[j] := - d[j];
   end;
 end;
end;
```

*Remark.* For the case $k = 2$ there is another solution that uses the binary system. (It is this solution that is usually associated with the name "Gray code".)

Let us write down all the numbers $0, \ldots, 2^n - 1$ in binary notation. For example, for $n = 3$ we have:

$$000 \quad 001 \quad 010 \quad 011 \quad 100 \quad 101 \quad 110 \quad 111$$

Each number is then transformed according to the following rule: each digit (except the first one) is replaced by its sum (modulo 2) with the preceding (untransformed) digit. In other words, the number with binary digits $a_1, a_2, \ldots, a_n$ is transformed into the number with binary digits $a_1, a_1 + a_2, a_2 + a_3, \ldots, a_{n-1} + a_n$ (addition modulo 2). For $n = 3$, we get the following list:

$$000 \quad 001 \quad 011 \quad 010 \quad 110 \quad 111 \quad 101 \quad 100$$

It is easy to check that the transformation described (which can be applied to any sequence of $n$ binary digits, giving another sequence of the same length) is invertible. Therefore, the list obtained contains all sequences of length $n$.

On the other hand, adding 1 to a number in binary notation means replacement of the suffix 011...1 by 100...0. This change leads to a change of exactly one digit after the transformation is applied. □

**Digression: An application of Gray codes.** Assume that some mechanical device has a rotating drum and we wish to get information about the rotation angle. If we make half the drum white, the remaining half black, and use a light sensor, we can measure the position of the drum up to 180°.

Drum cover:

If we make another track with black and white parts, and use a second light sensor, we can measure the position angle up to 90°:

$$
\begin{array}{cccc}
0 & 0 & 1 & 1 \\
0 & 1 & 0 & 1
\end{array}
$$

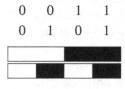

With a third track,

$$
\begin{array}{cccccccc}
0 & 0 & 0 & 0 & 1 & 1 & 1 & 1 \\
0 & 0 & 1 & 1 & 0 & 0 & 1 & 1 \\
0 & 1 & 0 & 1 & 0 & 1 & 0 & 1
\end{array}
$$

the precision becomes 45°, etc. However, there is a problem with this scheme. When two light sensors change their state from black to white, these changes may not happen at exactly the same time, and for a while the data are senseless.

We can use Gray codes to overcome this difficulty: we arrange the black and white sectors in such a way that only one track changes color each time. (This is also true for the last change after a complete rotation is performed.)

$$
\begin{array}{cccccccc}
0 & 0 & 0 & 0 & 1 & 1 & 1 & 1 \\
0 & 0 & 1 & 1 & 1 & 1 & 0 & 0 \\
0 & 1 & 1 & 0 & 0 & 1 & 1 & 0
\end{array}
$$

The above formula allows us to convert the sensor data into the corresponding rotation angle easily.

**2.5.3.** Generate all permutations of the numbers 1..n in such a way that each permutation is obtained from the preceding one by an exchange (transposition) of two adjacent numbers. For example, for n=3, one of the possible answers is

$$3.2\ 1 \to 2\ 3.1 \to 2.1\ 3 \to 1\ 2.3 \to 1.3\ 2 \to 3\ 1\ 2$$

(the dots indicate which numbers are exchanged at each step).

*Solution.* Put the set of all permutations into one-to-one correspondence with another set. This latter set contains all sequences y[1]..y[n] of nonnegative integers such that y[1] ⩽ 0, ..., y[n] ⩽ n-1. It has the same cardinality as the set of all permutations. The one-to-one correspondence is established as follows: Each permutation corresponds to the sequence y[1]..y[n], where y[i] is the number of j's such that both (a) j < i and (b) j is located to the left of i in this permutation. Why is it a one-to-one correspondence? Any permutation of 1..n can be obtained from a permutation of 1..n-1 by inserting n into one of the n places (before the first term, between the first and the second terms,..., after the last term). What does this insertion mean for the corresponding sequence of integers? A number that ranges from 0 to n-1 is appended to the end while the other terms remain unchanged.

This one-to-one correspondence can be explained by the following metaphor. Consider n cards with numbers 1..n written on the cards, and a growing pile made of the cards. Initially the pile has only one card with number 1 written on it. At the next step we add the card with number 2. There are two possible positions for that card (either before the first card or after it). Then we add the card with number 3 on it; there are three possible positions, etc. After we add the last card (there are n possible positions), we get a permutation of the numbers 1..n. This permutation is determined by positions chosen at steps 1..n; if we denote by y[i] the number of cards before the inserted card at step i, we get the one-to-one correspondence defined above.

We make one more remark about this correspondence. Assume that we increase or decrease y[i] by 1 for some i (leaving y[j] unchanged for all j ≠ i). Assume also that all subsequent y[j] (for all j > i) have maximal or minimal values. In this case two adjacent numbers in our permutation are exchanged. Namely, an increase in y[i] means that i is exchanged with its right neighbor, while a decrease means that i is exchanged with its left neighbor.

Recall how we generated all sequences of numbers 1..k in such a way that each sequence differs from the preceding sequence in one and only one place by using n × k rectangle. Now replace it by a board that resembles a staircase (the i-th column is a rectangle of width 1 and height i). Moving pieces according to the rules described above (using arrows on pieces), we traverse all the sequences, and the property mentioned above (i-th term changes only if all subsequent terms are maximal or minimal) holds.

To implement this scheme we need to modify the permutation according to the changes on the board. An obvious approach is to search for a given number i at each step. We can save ourselves some work if we keep (in addition to the permutation itself) the function

$$i \mapsto \text{position of i in the permutation;}$$

that is, the inverse mapping, and update both the permutation and its inverse. Here is the program:

```
program test;
  const n = ...;
  var
    x: array [1..n] of 1..n; {permutation}
    inv_x: array [1..n] of 1..n; {inverse permutation}
    y: array [1..n] of integer; {y[i] < i}
    d: array [1..n] of -1..1; {arrows}
    b: Boolean;

  procedure print_x;
    var i: integer;
  begin
    for i := 1 to n do begin
      write (x[i], ' ');
    end;
    writeln;
  end;

  procedure set_first; {first: y[i]=0 for all i}
    var i : integer;
  begin
    for i := 1 to n do begin
      x[i]  := n + 1 - i;
      inv_x[i]  := n + 1 - i;
      y[i]  := 0;
      d[i]  := 1;
    end;
  end;

  procedure move (var done : Boolean);
    var i, j, pos1, pos2, val1, val2, tmp : integer;
  begin
    i := n;
    while (i > 1) and (((d[i]=1) and (y[i]=i-1)) or
                ((d[i]=-1) and (y[i]=0))) do begin
```

```
      | i := i-1;
      end;
      done := (i > 1);
      {simplification: the first term cannot be changed}
      if done then begin
        y[i] := y[i] + d[i];
        for j := i+1 to n do begin
        | d[j] := -d[j];
        end;
        pos1 := inv_x[i];
        val1 := i;
        pos2 := pos1 + d[i];
        val2 := x[pos2];
        {pos1, pos2 are positions of elements to be
           exchanged; val1, val2 are its values; val2 < val1}
        tmp := x[pos1];
        x[pos1] := x[pos2];
        x[pos2] := tmp;
        tmp := inv_x[val1];
        inv_x[val1] := inv_x[val2];
        inv_x[val2] := tmp;
      end;
    end;

begin
  set_first;
  print_x;
  b := true;
  {all permutations up to the current one (including it)
   are printed;
   if b is false, the current one is the last one}
  while b do begin
    move (b);
    if b then print_x;
  end;
end.
```

## 2.6 Some remarks

Let us review the approach we've been using. We introduce some order on the objects
to be generated and write a procedure that obtains the next object (in this order). In
the Gray code problems, we were forced to maintain some additional information
(directions of arrows). Finally, when generating permutations in such a way that
only two numbers are exchanged at a time, we establish a one-to-one correspondence

between the set to be generated and some other (presumably simpler) set. There are some cases where this trick is useful. In this section, we consider several problems of this type connected with the so-called Catalan numbers.

**2.6.1.** Generate all sequences of length 2n, composed of 1s and −1s, satisfying the following conditions: (a) the sum of all terms is 0; (b) the sum of any prefix is nonnegative; that is, the number of −1s does not exceed the number of 1s. (The number of such sequences is called the *Catalan number*; see the formula for Catalan numbers on p. 48, problem 2.7.3.)

*Solution.* Represent 1 by a vector $(1,1)$ and represent −1 by $(1,-1)$. In terms of vectors, we are looking for all paths from $(0,0)$ to $(2n,0)$ that never go below the $x$-axis.

Let us generate the sequences in alphabetic order (assuming that −1 precedes 1). The first sequence is the "zig-zag"

$$1, -1, 1, -1, \ldots$$

The last sequence will be the sequence

$$1, 1, 1, \ldots, 1, -1, -1, \ldots, -1.$$

But how do we generate the next sequence? It should coincide with the current sequence up to some point where they differ and −1 is replaced by 1. This place should be as close to the end as possible. But there is a restriction; −1 may be replaced by 1 only if there is 1 on the right of it (which can be replaced by −1). After we replace −1 by 1, we are faced with the following problem: A prefix of the sequence is fixed; find the minimal sequence with that prefix. The solution: extend the given prefix step by step; at each step append −1 if possible (the sum must be nonnegative); otherwise, append 1. Here is the resulting program:

```
...
type array2n = array [1..2n] of integer;
...
procedure get_next (var a: array2n; var last: Boolean);
  {a is replaced by the next sequence if it exists
   (and last:=false), otherwise last:=true}
  var k, i, sum: integer;
begin
  k:=2*n;
  {invariant: a[k+1..2n] contains only -1s}
  while a[k] = -1 do begin k:=k-1; end;
  {k is maximal among all k such that a[k]=1}
  while (k>0) and (a[k] = 1) do begin k:=k-1; end;
  {a[k] is the rightmost -1 preceding some 1;
   k=0 if there is no -1 on the left of 1}
  if k = 0 then begin
    last := true;
  end else begin
```

```
    last := false;
    i:=0; sum:=0;
    {sum = a[1]+...+a[i]}
    while i<>k do begin
    | i:=i+1; sum:= sum+a[i];
    end;
    {sum = a[1]+...+a[k], a[k]=-1}
      a[k]:= 1; sum:= sum+2;
    {all a[1]..a[k] have their final values,
       sum=a[1]+...+a[k]}
    while k <> 2*n do begin
      k:=k+1;
      if sum > 0 then begin
      | a[k]:=-1
      end else begin
      | a[k]:=1;
      end;
      sum:= sum+a[k];
    end;
    {k=2n, sum=a[1]+...a[2n]=0}
  end;
end;                                                              □
```

**2.6.2.** Find all possible ways to compute the product of n factors. (The order of the factors remains unchanged.) Each multiplication should be indicated by parentheses. For example, for n = 4, the following five expressions should be generated:

$$((ab)c)d, \quad (a(bc))d, \quad (ab)(cd), \quad a((bc)d), \quad a(b(cd)).$$

[*Hint*. Each order of operations corresponds to a sequence of commands of the stack calculator described on p. 127.]                                   □

**2.6.3.** There are 2n points on a circle numbered (along the circle) by the numbers 1..2n. Generate all possible ways to draw n non-intersecting segments having those 2n points as endpoints.                                              □

**2.6.4.** Generate all ways to cut a convex polygon with n vertices into triangles using n−2 diagonals.                                                       □

(We will discuss polygon triangulations in chapter 8 on dynamic programming, p. 121.)

## 2.7 Counting

In this chapter we considered several methods that may be used to generate all the elements of a given finite set. One more approach will be considered below (under the name of "backtracking") in chapter 3. But sometimes it is much easier to count all the objects with some property than it is to generate them. The classic example is $\binom{n}{k}$, which is the number of $k$-element subsets of an $n$-element set. These numbers form the "Pascal triangle" and can be computed using the identities

$$\binom{n}{0} = \binom{n}{n} = 1 \qquad (n \geqslant 1)$$

$$\binom{n}{k} = \binom{n-1}{k-1} + \binom{n-1}{k} \qquad (n > 1, \ 0 < k < n)$$

or the formula

$$\binom{n}{k} = \frac{n!}{k! \cdot (n-k)!}.$$

(The first method is more efficient when many values of $\binom{n}{k}$ for different $n$ and $k$ are needed.)

Let us give some other examples.

**2.7.1.** (Number of partitions) Let $P(n)$ be the number of representations of a nonnegative integer $n$ as a sum of positive integer summands (order is insignificant; that is, the representations $1 + 2$ and $2 + 1$ are identical). We assume that $P(0) = 1$ (the only representation has no summands at all). Write a program that finds $P(n)$ for a given $n$.

*Solution.* One can prove the following (nontrivial) formula for $P(n)$:

$$P(n) = P(n-1) + P(n-2) - P(n-5) - P(n-7) + P(n-12) + P(n-15) + \cdots$$

(terms are grouped in pairs, the signs before the pairs alternate, arguments in $q$-th pair are $n - (3q^2 - q)/2$ and $n - (3q^2 + q)/2$). We assume $P(k) = 0$ for $k \leqslant 0$, so the sum is finite.

Even if we did not know this formula, there is a way to compute $P(n)$ that is much more efficient than counting all the partitions one-by-one.

By $R(n, k)$ (defined for $n \geqslant 0, k \geqslant 0$) we denote the number of representations of $n$ as a sum of positive integers not exceeding $k$. Let $R(0, k)$ be equal to 1 for all $k \geqslant 0$. Evidently, $P(n) = R(n, n)$. All the representations of $n$ are classified according to the maximal summand (which is denoted by $i$ in the sequel). The number $R(n, k)$ is the sum over all $i$ in $\{1, \ldots, k\}$ of the number of partitions with elements not exceeding $k$ and maximal element $i$. The partitions of $n$ into a sum where all terms do not exceed $k$ and maximal term is equal to $i$ are in one-to-one correspondence with the partitions of $n - i$ into terms not exceeding $i$ (assuming that $i \leqslant k$). Therefore,

$$R(n, k) = \sum_{i=1}^{k} R(n - i, i) \quad \text{for } k \leqslant n;$$

$$R(n, k) = R(n, n) \quad \text{for } k > n.$$

These equations allows us to construct a table of values of the function $R$.  $\square$

**2.7.2.** (Lucky numbers) A sequence of $2n$ digits (each digit is in the $0, \ldots, 9$ range) is called "lucky" if the sum of the first $n$ digits is equal to the sum of the last $n$ digits. Find the number of all lucky sequences of a given length.

*Solution.* Let us generalize the problem and find the number $T(n, k)$ of sequences of length $2n$ where the difference between the sum of first $n$ digits and the sum of the last $n$ digits is equal to $k$ (where $-9n \leqslant k \leqslant 9n$).

We divide all these sequences into classes according to the difference between the first and last digit. If this difference is equal to $t$, the difference between the remaining sums of $n - 1$ digits is $k - t$. Note that there are $10 - |t|$ pairs of decimal digits with difference $t$. So we get the formula:

$$T(n, k) = \sum_{t=-9}^{9} (10 - |t|) T(n - 1, k - t).$$

(Some terms may be missing if $k - t$ is too large.)                                    □

In some cases, the answer may be given by an explicit formula. For example, this is the case for Catalan numbers.

**2.7.3.** Prove that the Catalan number, i.e., the number of sequences of length $2n$ composed of $n$ ones and $n$ minus ones such that each initial segment has a non-negative sum, is equal to $\binom{2n}{n}/(n + 1)$.

[*Hint.* The Catalan number is the number of polygonal paths going from $(0, 0)$ to $(2n, 0)$ formed by vectors $(1, 1)$ and $(1, -1)$ that do not intersect the half-plane $y < 0$. Therefore, this number is the difference between the number of all polygonal paths of the type described (which is $\binom{2n}{n}$) and the number of paths that intersect the half-plane $y < 0$. All paths of the type described that intersect the half-plane $y < 0$ intersect the line $y = -1$. If we reflect the part of the polygonal path that is on the right of the rightmost intersection point, we get a one-to-one correspondence between the polygonal paths in question and all polygonal paths from $(0, 0)$ to $(2n, -2)$. It remains to check that $\binom{2n}{n} - \binom{2n}{n+1} = \binom{2n}{n}/(n + 1)$.]                          □

# 3

# Tree traversal (backtracking)

In the preceding chapter we considered several problems that required us to enumerate all elements of some set $X$. The solution used the following scheme: A linear ordering on $X$ was imposed and a procedure to generate the next element of $X$ (according to that order) was described.

Sometimes this scheme cannot be applied directly. In this chapter, we consider another useful approach that allows us to generate all elements of some set. It is called "backtracking" or "tree traversal".

## 3.1 Queens not attacking each other: position tree

This approach is fairly general; however, we prefer to start with a specific example.

**3.1.1.** Generate all the positions of $n$ queens on an $n \times n$ chess board such that the queens are not attacking each other.

*Solution.* Evidently, each of $n$ rows should contain exactly one queen. By $k$-position we mean a position where $k$ queens occupy $k$ rows (starting from the bottom of the chess board) containing exactly one queen each. We do not impose any restrictions as yet and we allow positions where some queens are attacking other queens.

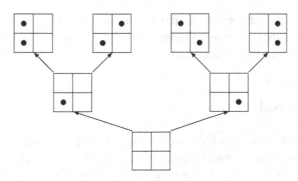

A. Shen, *Algorithms and Programming*, Springer Undergraduate Texts
in Mathematics and Technology, DOI 10.1007/978-1-4419-1748-5_3,
© Springer Science+Business Media, LLC 2010

Arrange all positions into a tree, whose root is the empty position ($k = 0$). Each $k$-position has exactly $n$ descendants, which have an additional queen in the $(k + 1)$-th row (in one of the columns $1, \ldots, n$). These $n$ descendants are ordered from left to right according to the position of the last (i.e., the uppermost) queen.

We are to select (among the vertices of this tree) those $n$-positions where queens are not attacking each other. To find them, our program will traverse the positions tree. To avoid unnecessary work, we make use of the following fact: If some tree vertex corresponds to a position where queens are attacking each other, all descendants of this vertex have the same property and therefore may be ignored safely. Therefore, this part of the position tree may be discarded.

Let us give some relevant definitions. A $k$-position is called "admissible" if *after the $k$-th queen is removed*, the remaining queens are not attacking each other. Our program will consider only admissible positions.

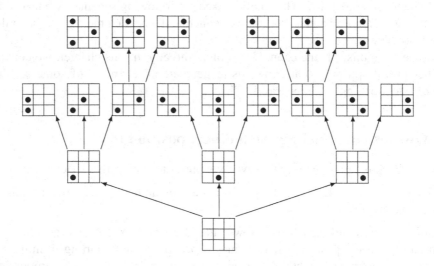

The tree of admissible positions for $n = 3$

Now the queens problem can be divided in two parts: (1) how to traverse all the vertices of a given tree; (2) how to represent the tree of admissible positions for the queens problem using Pascal constructs.

## 3.2 Tree traversal

Let us formulate the general problem of visiting all the vertices of a given tree. Imagine there is a robot that can be placed at any vertex of a tree. (Vertices are shown as small circles in our pictures.) The repertoire of the robot consists of the following commands:

- up_left   ("move along the up-left arrow")
- right   ("move to the right neighbor")
- down   ("move down one level")

(The pictures below show which movements correspond to these commands.)

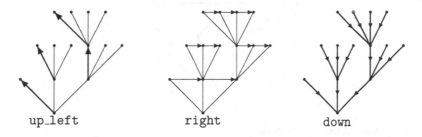

up_left               right               down

Moreover, the robot's repertoire includes tests that check whether each command can be executed:

- is_up;
- is_right;
- is_down

(the last test returns True everywhere except at the root). Please note that the right command allows a move from the vertex to its "brother" but not to its "cousin" having only a grandfather in common.

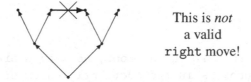

This is *not*
a valid
right move!

    Finally, we assume that the robot is able to perform a command process. Our goal is to process (that is, to execute the command process for) all leaves of the tree. (A *leaf* is a vertex such that is_up is false; that is, a vertex with no descendants.) In our chess problem, process means to check the position and to print it (if it contains n queens not attacking each other).

    *Remark.* Our trees (like most of the real trees) have root at the bottom and leaves at the top. Please be warned that in most computer science books trees are drawn with the root at the *top*. While it seems to be nonintuitive, it is the *de facto* standard.

    The proof of the program below uses the following conventions. Assume that the position of the robot is fixed. Then all the leaves of the tree are divided into three categories: (1) leaves *above* the robot; (2) leaves *on the left* of the robot and (3) leaves *on the right* of the robot. Indeed, the (unique) path from the root to a given leaf (a) may go through the robot's position; (b) may turn to the left before the robot's position, or (c) may go to the right before it. By (LP) we denote the condition "all the leaves on the left of the robot are processed"; by (LAP) we denote the condition "all the leaves on the left of the robot and above it are processed". (In both cases we require that no other leaves are processed.)

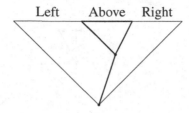

Left    Above    Right

We will use the following procedure:

```
procedure go_up_and_process;
| {before: (LP), after: (LAP)}
begin
  | {invariant: LP}
  | while is_up do begin
  |   | up_left;
  | end
  | {LP, current position is a leaf}
  | process;
  | {LAP}
end;
```

Here is the main program:

```
before: robot is in the root, no leaves are processed
after: robot is in the root, all leaves are processed

{LP}
go_up_and_process;
{invariant: LAP}
while is_down do begin
  | if is_right then begin {LAP, is_right}
  |   | right;
  |   | {LP}
  |   | go_up_and_process;
  | end else begin
  |   | {LAP, not is_right, is_down}
  |   | down;
  | end;
end;
{LAP, current position is root=>all leaves are processed}
```

Correctness now follows from the properties of the robot's commands. They are presented below in the format:

$$\{\text{precondition}\} \text{ command } \{\text{postcondition}\};$$

The postcondition is guaranteed after execution of the command, assuming that the precondition was true before:

(1)  {LP, not is_up} process {LAP}

(2)  {LP} up_left {LP}

(3)  {is_right, LAP} right {LP}

(4)  {not is_right, is_down, LAP} down {LAP}

These properties follow directly from the definitions. Indeed, if we are in the leaf, there is only one leaf above the robot, so processing it converts LP to LAP, according to (1). For (2), we note that is_up does not change the set of leaves on the left of current position. To prove (3), we note that the leaves on the left of the right brother of the current position, are the leaves on the left and above the current position. Finally, down does not change the set of leaves above and on the left of current position (just moving some leaves from the second category to the first one).

**3.2.1.** Prove that the program shown above terminates for any finite tree.

*Solution.* The procedure go_up_and_process terminates (since the height of the robot position cannot increase indefinitely). Assume that the program as a whole does not terminate. Leaves are never processed twice and the number of leaves is finite. Therefore, there is a moment after which leaves are not processed. This is possible only if the robot goes down at each step, but this is a contradiction. (The estimate for the number of operations will be given later.)  □

**3.2.2.** Prove that the following program also processes all the leaves of a tree (once each):

```
var state: (LP, LAP);
state := LP;
while is_down or (state <> LAP) do begin
  if (state = LP) and is_up then begin
  | up_left;
  end else if (state = LP) and not is_up then begin
  | process; state := LAP;
  end else if (state = LAP) and is_right then begin
  |   right; state := LP;
  end else begin {state = LAP, not is_right, is_down}
  |   down;
  end;
end;
```

*Solution.* The invariant relation: The value stored in the variable state is correct; that is,

$$\text{state} = \text{LP} \quad \Rightarrow \quad \text{LP is true}$$
$$\text{state} = \text{LAP} \quad \Rightarrow \quad \text{LAP is true}$$

The proof of termination: the change from LP to LAP is possible only when a vertex is processed. Therefore, if the program does not terminate, the variable `state` achieves its final value and does not change further, which is impossible.          □

**3.2.3.** Write a program that traverses the tree and processes all vertices (not only leaves).

*Solution.* Let $x$ be a vertex. Then all vertices of the tree can be divided into four categories. Indeed, let $y$ be some other vertex. Consider the path from the root to $y$. Four cases are possible:

(a) this path is a prefix of the path from the root to $x$ ($y$ is *below* $x$);
(b) this path turns to the left before reaching $x$ ($y$ is *on the left* of $x$);
(c) this path goes through $x$ ($y$ is *above* $x$);
(d) this path turns to the right before reaching $x$ ($y$ is *on the right* of $x$).

In particular, the vertex $x$ belongs to class (c). Now the following conditions are used in our program:

(ULP) all vertices under the current position and on the left of it are processed;

(ULAP) all vertices under the current position, on the left of it, and above it are processed.

Here is the program:

```
procedure go_up_and_process;
  {before: (ULP), after: (ULAP)}
begin
  {invariant: ULP}
  while is_up do begin
    process;
    up_left;
  end
  {ULP, the current position is a leaf}
  process;
  {ULAP}
end;
```

The main algorithm:

```
before: robot is in the root, no vertices are processed
after: robot is in the root, all vertices are processed

{ULP}
go_up_and_process;
{invariant: ULAP}
while is_down do begin
```

```
    if is_right then begin {ULAP, is_right}
     |  right;
     |  {ULP}
     |  go_up_and_process;
    end else begin
     |  {ULP, not is_right, is_down}
     |  down;
    end;
  end;
  {ULAP, robot in the root => all vertices processed}          □
```

**3.2.4.** The program given in the solution of the preceding problem processes any vertex before its descendants. Modify the program in such a way that any non-leaf vertex will be processed twice, once *before* and once *after* its descendants. (The leaves should be processed once.)

*Solution.* In the program below, by "Under-Left-Processed" (ULP) we mean "all the vertices under the current position of the robot are processed once; all the vertices on the left are processed completely" (that is, leaves are processed once, all other vertices are processed twice: once before and once after their descendants). By "Under-Left-Above-Processed" (ULAP) we mean "all the vertices under the current position of the robot are processed once; all vertices on the left of and above the current position are processed completely".

Here is the auxiliary procedure:

```
    procedure go_up_and_process;
     |  {before: (ULP), after: (ULAP)}
    begin
     |  {invariant: ULP}
     |  while is_up do begin
     |   |  process;
     |   |  up_left;
     |  end
     |  {ULP, the current position is a leaf}
     |  process;
     |  {ULAP}
    end;
```

The main program:

```
    before: robot is in the root, no vertices are processed
    after: robot is in the root, all vertices are processed

    {ULP}
    go_up_and_process;
    {invariant: ULAP}
    while is_down do begin
```

```
    if is_right then begin {ULAP, is_right}
      right;
      {ULP}
      go_up_and_process;
    end else begin
      {ULP, not is_right, is_down}
      down;
      process;
    end;
  end;
  {ULAP, robot is in the root =>
      all vertices are processed completely}                  □
```

**3.2.5.** Prove that the number of operations in this program is proportional to the number of vertices. (Therefore, the same is true for the programs given above that differ from the last one only because some process commands have been omitted.)

[*Hint*. Roughly speaking, each second operation is processing some vertex, and any vertex is processed at most twice.]                                          □

## 3.3 Queens: position tree implementation

Let us return to the queens problem. In this problem, we use only the first and simplest of our tree traversal programs, which processes each leaf once.

We implement all the operations for the case of the positions tree. Each position is represented by a variable $k:0..n$ (the number of queens) and an array $c: array[1..n]$ of $1..n$. Here $c[i]$ is the horizontal coordinate of the $i$-th queen (whose vertical coordinate is $i$). If $i > k$, the value of $c[i]$ is insignificant. Only the admissible positions are included in the tree. (According to our definition, a position is admissible if after the uppermost queen is removed, no queens are attacking each other.)

Now we are ready to present the program that solves the queens' problem:

```
program queens;
  const n = ...;
  var
    k: 0..n;
    c: array [1..n] of 1..n;

  procedure begin_work; {initialize}
  begin
    k := 0;
  end;
```

```
function danger: Boolean;
 | {the uppermost queen is under attack}
 | var b: Boolean; i: integer;
begin
 | if k <= 1 then begin
 |  | danger := false;
 | end else begin
 |  | b := false; i := 1;
 |  | {b <=> the uppermost queen is under attack of
 |  |    some queen with y-coordinate < i}
 |  | while i <> k do begin
 |  |  | b := b or (c[i]=c[k]) {vertical}
 |  |  |     or (abs(c[i]-c[k])=abs(i-k)); {diagonal}
 |  |  | i := i+1;
 |  | end;
 |  | danger := b;
 | end;
end;

function is_up: Boolean;
begin
 | is_up := (k < n) and not danger;
end;

function is_right: Boolean;
begin
 | is_right := (k > 0) and (c[k] < n);
end;
{danger: when k=0, the value c[k] is undefined}

function is_down: Boolean;
begin
 | is_down := (k > 0);
end;

procedure up_left;
begin {k < n, not danger}
 | k := k + 1;
 | c [k] := 1;
end;

procedure right;
begin {k > 0,  c[k] < n}
 | c [k] := c [k] + 1;
```

```
      end;

      procedure down;
      begin {k > 0}
      | k := k - 1;
      end;

      procedure process;
      | var i: integer;
      begin
      | if (k = n) and not danger then begin
      |   for i := 1 to n do begin
      |   | write ('<', i, ',' , c[i], '> ');
      |   end;
      |   writeln;
      | end;
      end;

      procedure go_up_and_process;
      begin
      | while is_up do begin
      | | up_left;
      | end;
      | process;
      end;

    begin
    | begin_work;
    | go_up_and_process;
    | while is_down do begin
    |   if is_right then begin
    |   | right;
    |   | go_up_and_process;
    |   end else begin
    |   | down;
    |   end;
    | end;
    end.
```
                                                                    □

**3.3.1.** The program above spends a lot of time inside the procedure is_up (to check if the uppermost queen is under attack, we need $O(n)$ operations). Modify the implementation of the positions tree in such a way that all three tests is_up/is_right/is_down and the corresponding three commands require only $O(1)$ operations (that is, the number of operations for any of them should be limited by a constant that does not depend on n).

*Solution.* For any vertical and for any diagonal line (there are two types of diagonal lines — ascending and descending ones) let us introduce a Boolean variable that
indicates if this line is occupied by some queen (except the uppermost one, which is
ignored). Note that any of those lines may be occupied by at most one queen (because
the position is assumed to be admissible).                                                    □

## 3.4 Backtracking in other problems

**3.4.1.** Use backtracking in the following problem: An array of n positive integers a[1]..a[n] and a positive integer s are given. Determine if s can be represented as a sum of some of the elements of the array a. (Each element may be used
at most once.)

*Solution.* Construct the position tree as follows: The k-*position* is a sequence of
k Boolean values that determines which of the elements a[1]..a[k] are used as
summands. The position is *admissible* if the sum of the corresponding elements does
not exceed s.                                                                                  □

*Remark.* This approach is better than an exhaustive search that considers all $2^n$
subsets. We may also sort the array a in descending order. Also, we can change the
definition of an admissible position to exclude positions where the sum of rejected
elements is larger than the difference between s and the sum of all accepted elements.
However, this does not lead to a fundamental improvement; this problem belongs to
the category of the so-called "NP-complete problems". See the book by A. Aho
J. Hopcroft and J. Ullman [1] and the book by M.R. Garey and D.S. Johnson [6].

This problem is traditionally called "the knapsack problem": A knapsack that
is capable of carrying s pounds should be filled completely using only objects of
weights a[1]..a[n]. See the last problem of section 8.1 (p. 123), where a "dynamic
programming" algorithm is given whose running time is polynomial in n + s.

**3.4.2.** Generate all sequences of *n* digits 0, 1 and 2 that do not contain a substring
of type *XX*. (E.g., the sequence 210102 is prohibited because it contains 1010.)    □

**3.4.3.** Repeat the previous problem for binary strings of length *n* that do not
contain a substring of type *XXX*.                                                            □

Another problem of the same category: "Is it possible to compose a given polygon of 'pentamino' blocks?" The crucial component of an effective algorithm for
such a problem is a good criterion that can (in some cases) guarantee that a given
position cannot be extended to a solution of the problem and therefore may be
discarded.

# 4

# Sorting

Sorting is a simple and practically important example of advantages provided by efficient algorithms (over straightforward ones). First (section 4.1) we formulate the sorting problem and show two straightforward (but inefficient) algorithms. Then (section 4.2) we show two much more efficient algorithms (merge sort and heap sort) that have running times proportional to $n \log n$ (instead of $n^2$). In section 4.3 we show how sorting algorithms can be applied even if the statement of the problem does not mention sorting. Section 4.4 provides some lower bound for the number of comparisons needed for sorting (thus showing that our algorithm cannot be significantly improved). Finally, in section 4.5 we consider some nice problems related to sorting.

## 4.1 Quadratic algorithms

**4.1.1.** Let $a[1], \ldots, a[n]$ be an array of numbers (say, integers). Construct the array $b[1], \ldots, b[n]$ that contains the same numbers in increasing order: $b[1] \leqslant \ldots \leqslant b[n]$.

*Remark.* The elements $a[1] \ldots a[n]$ need not be distinct. In this case we require that the multiplicity (=number of occurrences) of each number in $b[1] \ldots b[n]$ should be equal to its multiplicity in $a[1] \ldots a[n]$.

*Solution.* It is convenient to consider $a[1] \ldots a[n]$ and $b[1] \ldots b[n]$ as the initial and final values of some array $x$. The requirement "a and b contain the same numbers" will be guaranteed if the only operation permitted on $x$ is the exchange of two its elements. (Of course, we are also allowed to read elements of $x$.)

```
k := 0;
{k minimal elements of x are in their places}
while k <> n do begin
  s := k + 1; t := k + 1;
  {x[s] is minimal among x[k+1]...x[t] }
```

A. Shen, *Algorithms and Programming*, Springer Undergraduate Texts in Mathematics and Technology, DOI 10.1007/978-1-4419-1748-5_4, © Springer Science+Business Media, LLC 2010

```
while t<>n do begin
    t := t + 1;
    if x[t] < x[s] then begin
        s := t;
    end;
end;
{x[s] is minimal among x[k+1]..x[n] }
... exchange x[s] and x[k+1];
k := k + 1;
end;
```

□

**4.1.2.** Give another sorting algorithm which uses the following invariant relation: "first k elements are sorted" ($x[1] \leqslant \ldots \leqslant x[k]$).

*Solution.* (This algorithm is called *insertion sort.*)

```
k:=1;
{first k elements are sorted}
while k <> n do begin
    t := k + 1;
    {k+1-th element moves to the left until it finds its
     place; t is its current position}
    while (t > 1) and (x[t] < x[t-1]) do begin
        ... exchange x[t-1] and x[t];
        t := t - 1;
    end;
end;
```

*Remark.* Danger: When (t > 1) is false, the test x[t] < x[t-1] refers to a non-existing value x[0].                                                                      □

Both of the above solutions require a number of operations proportional to $n^2$. There are more efficient algorithms, however, as we shall see.

## 4.2 Sorting in $n \log n$ operations

**4.2.1.** Find a sorting algorithm that requires only $O(n \log n)$ operations. (In other words, the number of operations should not exceed $Cn \log n$ for some constant $C$ that does not depend on $n$.)

We give two solutions for this problem.

*Solution* 1 (merge sort.)
Let k be a positive integer, and split the array x[1]..x[n] into segments of length k. (The first segment is x[1]..x[k], the next is the segment x[k+1]..x[2k], etc.) The last segment is incomplete if n is not a multiple of k. We say that the array

x is k-*sorted* if each of these segments (considered separately) is sorted. Of course, any array is 1-sorted. If an array of length n is k-sorted for k $\geqslant$ n, it is sorted.

Assume there is a procedure that transforms any k-sorted array into a 2k-sorted array (containing the same elements). Using this procedure, we write down our sorting algorithm as follows:

```
k:=1;
{the array x is k-sorted}
while k < n do begin
    ..transform the k-sorted array into a 2k-sorted array;
    k := 2 * k;
end;
```

How do we construct such a procedure? It repeats the following step: two sorted segments of length at most k are merged into one sorted segment. Assume that the procedure

merge (p,q,r: integer)

called with p $\leqslant$ q $\leqslant$ r merges two already sorted segments x[p+1]..x[q] and x[q+1]..x[r] into a sorted segment x[p+1]..x[r] (without changing other parts of the array x):

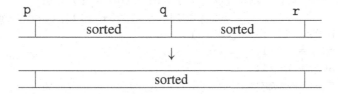

The transformation of a k-sorted array into a 2k-sorted array then goes as follows:

```
t:=0;
{t is a multiple of 2k or t = n, x[1]..x[t] is
 2k-sorted; the rest of x is unchanged}
while t + k < n do begin
    p := t;
    q := t+k;
    ...r := min (t+2*k, n);
            {min(a,b) is the minimum of a and b}
    merge (p,q,r);
    t := r;
end;
```

The merge procedure uses an auxiliary array as temporary storage for the result. This auxiliary array will be denoted by b. Let p0 and q0 be the indices of the last elements merged; s0 is the index of the last element written to b. At each step, one of the two following actions is performed:

```
b[s0+1]:=x[p0+1];
p0:=p0+1;
s0:=s0+1;
```

or

```
b[s0+1]:=x[q0+1];
q0:=q0+1;
s0:=s0+1;
```

(C fans will enjoy the shorthands b[++s0]=x[++p0] and b[++s0]=x[++q0] here.)

The first action (where the element is taken from the first segment) may be performed if the following two conditions are fulfilled:

(1) the first segment is not empty (p0 < q); and

(2) the second segment is empty (q0 = r) or its first element is greater than or equal to the first element of the first segment   [(q0 < r) and (x[p0+1] ≤ x[q0+1])].

The conditions that make the second action possible are similar. We obtain the following program:

```
p0 := p; q0 := q; s0 := p;
while (p0 <> q) or (q0 <> r) do begin
  if (p0 < q) and ((q0 = r) or ((q0 < r) and
                     (x[p0+1] <= x[q0+1]))) then begin
    b [s0+1] := x [p0+1];
    p0 := p0+1;
    s0 := s0+1;
  end else begin
    {(q0 < r) and ((p0 = q) or ((p0<q) and
       (x[p0+1] >= x[q0+1])))}
    b [s0+1] := x [q0+1];
    q0 := q0 + 1;
    s0 := s0 + 1;
  end;
end;
```

(If both segments are nonempty and have equal first elements, both actions are legal. In this case, the program chooses the first one.)

The only thing left to do is copy the merged array back into x. (Warning: If you decide to perform copying outside the merge procedure, please note that the last segment should be copied even it was not merged with anything.)

The program has a standard deficiency: the array index can be out of bounds when the Boolean expressions are evaluated (if "short-circuit evaluation" is not used).

It is easy to see that the number of operation needed to transform a $k$-sorted array into a $2k$-sorted one is proportional to $n$ (each element requires a bounded number of operations), We need about $\log n$ passes to increase $k$ from 1 to $n$, so the total number of operations in $O(n \log n)$ (=does not exceed $Cn \log n$ for some $C$ and all $n$).

*Solution* 2 (heap sort).

Draw a "complete binary tree". The root of this tree is drawn as a small circle at the bottom; two arrows go from the root node to the two nodes above it, two arrows go from each of them, etc.

We say that arrows connect a father to its two sons. Each node has two sons and one father unless it is the root (a node at the bottom) or a leaf (a node at the top). For simplicity, we assume for a moment that the length of the array to be sorted is a power of 2 and the elements completely fill some level of the tree. Fill the part of the tree below them using the following rule:

$$\text{father} = \min(\text{son}_1, \text{son}_2)$$

According to this rule, the value at the root of the tree will be the minimal element of the whole array.

Take the minimal element out of the array. To do that, we first locate it. It can be traced going from bottom to top, traversing the son that has the same value as its father. After the minimal element is removed, we replace it by the symbol $\infty$ and modify its ancestors going from top to bottom. (We assume that $\min(t, \infty) = t$.) Consequently, the root of the tree contains the second minimal element. We locate it, take it out (replacing it by $\infty$) and modify the tree. This procedure is repeated until all the elements are taken out and the root of the tree is occupied by $\infty$.

To write down the corresponding program the following convention is useful. Assume that the vertices of the tree are numbered by $1, 2, \ldots$ in such a way that root has number 1 and position n has sons 2n and 2n + 1. We do not give the details, because we will present a more efficient algorithm that does not use any additional memory (except for a fixed number of variables and the array itself). Here it is:

The elements to be sorted are placed at all levels of the tree, not just the upper level. Suppose we want to sort the array x[1]..x[n]. The tree has numbers 1..n as vertices. We assume that x[i] is placed at vertex i. During execution, the number of vertices in the tree will decrease. The current number of tree vertices is stored in k. Therefore, at any time the array x[1]..x[n] is divided into two parts. Its initial segment x[1]..x[k] represents a tree. The remaining part x[k+1]..x[n] contains the already sorted part of the array; those elements have already reached their final destination.

At each step, the algorithm extracts the maximal element from the tree and puts it into the sorted part (using the position freed when the tree becomes smaller).

Let us specify some terminology. The *vertices* of the tree are numbers from 1 up to the current value of k. Each vertex s may have *sons* 2s and 2s + 1. If both numbers are larger than k, the vertex s has no sons. Such a vertex is called a *leaf*. If 2s = k, the vertex s has exactly one son (2s).

For any s in 1..k, we consider a *subtree rooted at* s (or s-*subtree*). It contains the vertex s and all its descendants (sons, grandsons, etc. — until we leave the segment 1..k). The vertex s is called *regular* if the element placed in it is the maximal element of the s-subtree; the s-subtree is called *regular* if all its vertices are regular. (In particular, any leaf is a regular singleton subtree.) Please note that the validity of the statement "s-subtree is regular" depends not only upon s but also upon the current value of k.)

*Remark.* Modern textbooks (see, e.g., [3]) use terms "child" ("parent", "sibling", etc.) instead of "son" ("father", "brother", etc.) that are used in older textbooks (see, e.g., [1]). When using this new terminology, you should have in mind that each vertex has only one parent, only one grandparent, etc.

The general structure of the algorithm is as follows:

```
k:= n;
.. Make the 1-subtree regular;
{x[1],..,x[k] <= x[k+1] <=..<= x[n]; 1-subtree is
  regular, therefore, x[1] is maximal among x[1]..x[k]}
while k <> 1 do begin
    .. exchange x[1] and x[k];
    k := k - 1;
    {x[1]..x[k-1] <= x[k] <=...<= x[n]; 1-subtree is
      regular everywhere except the root (may be)}
    .. restore the regularity of 1-subtree
end;
```

As a tool, we use a procedure that restores the regularity of the subtree which is regular everywhere except its root. Here it is:

```
{s-subtree is regular everywhere except perhaps its root}
t := s;
{s-subtree is regular everywhere except perhaps t}
while ((2*t+1 <= k) and (x[2*t+1] > x[t])) or
        ((2*t <= k) and (x[2*t] > x[t])) do begin
    if (2*t+1 <= k) and (x[2*t+1] >= x[2*t]) then begin
        ... exchange x[t] and x[2*t+1];
        t := 2*t + 1;
    end else begin
        ... exchange x[t] and x[2*t];
        t := 2*t;
    end;
end;
```

Let us look closely at this procedure to check its correctness. Assume that all vertices of the s-subtree are regular except perhaps the vertex t. Consider the sons of t. They are regular and therefore contain maximal elements of their subtrees. Therefore, we have only three possibilities for the maximal element of the t-subtree, namely, the vertex t and its sons. If the vertex t contains the maximal element, this vertex is regular, and we are done. The while-construct can be rewritten as follows:

```
while the maximal element is not t but one of its sons
    do begin
 if it is the right son then begin
 | exchange t and its right son; t:= right son;
 end else begin {maximal element = the left son of t}
 | exchange t and its left son; t:= left son;
 end
end
```

After the exchange, the vertex t becomes regular (since it contains the maximal element of the t-subtree). The son that does not take part in the exchange is still regular. The other son may become irregular. Any other vertex u of the s-subtree remains regular because the value placed in u is unchanged, as well as the u-subtree (though elements of the subtree may be permuted).

The same procedure may be used at the first stage of our algorithm to make the 1-subtree regular:

```
k := n;
u := n;
{s-subtrees are regular for all s>u}
while u<>0 do begin
 {u-subtree is regular everywhere
             except the root (may be)}
 .. restore the regularity of u-subtree;
 u:=u-1;
end;
```

This algorithm is called *heap sort*.

It is easy to see that restoring regularity in a subtree (as described) requires $O(\log n)$ operations (each iteration moves us up in the tree). We need at most $O(n)$ restoring operations, so the total number of operations is $O(n \log n)$.

Now we are ready to write down the heapsort program in Pascal. We assume that n is a constant and x is a variable of type arr = array [1..n] of integer).

```
procedure sort (var x: arr);
 var u, k: integer;
 procedure exchange(i, j: integer);
 | var tmp: integer;
 | begin
 | tmp   := x[i];
```

```
      x[i] := x[j];
      x[j] := tmp;
    end;
  procedure restore (s: integer);
    var t: integer;
    begin
    t:=s;
    while ((2*t+1 <= k) and (x[2*t+1] > x[t])) or
          ((2*t <= k) and (x[2*t] > x[t])) do begin
      if (2*t+1 <= k) and (x[2*t+1] >= x[2*t]) then begin
        exchange (t, 2*t+1);
        t := 2*t+1;
      end else begin
        exchange (t, 2*t);
        t := 2*t;
      end;
    end;
  end;
begin
  k:=n;
  u:=n;
  while u <> 0 do begin
    restore (u);
    u := u - 1;
  end;
  while k <> 1 do begin
    exchange (1, k);
    k := k - 1;
    restore (1);
  end;
end;
```

Several remarks:

- The method used in the heapsort algorithm has other applications. One example is the priority queue implementation, see section 6.4, p. 100.
- The advantage of the merge sorting algorithm is that it does not require the entire array to be placed into RAM. We can sort portions of the array that fit into available RAM (say, using heapsort) and then merge the files obtained.
- Another important sorting algorithm ("Hoare quicksort") uses the following approach. To sort an array, choose a random element $b$ and split the array into three parts: elements smaller than $b$, equal to $b$ and greater than $b$. (This problem is discussed in chapter 1.) Now it remains to sort the first and the third parts, which can be done recursively using the same method. The number of steps of this algorithm is a random variable. One can prove that its expectation does not

exceed $Cn \log n$. It is one of the fastest algorithms in practice. (We shall discuss its recursive and non-recursive implementations later.)

- Finally, let us mention that sorting in $Cn \log n$ operations may be performed using the technique of balanced trees (see chapter 14), but the programs are rather complicated and the constant $C$ is large enough to make this method impractical. □

## 4.3 Applications of sorting

**4.3.1.** Find the number of different elements in an integer array. The number of operations should be of order $n \log n$. (This problem was already mentioned in chapter 1.)

*Solution.* Sort the array and then count the different elements going from left to right. □

**4.3.2.** Suppose that n closed intervals [a[i], b[i]] on the real line are given (i = 1..n). Find the maximal k such that there exists a point covered by k intervals (the maximal number of "layers"). The number of operations should be of order n log n.

[*Hint.* Sort all the left and right endpoints of the intervals together. While sorting, assume that the left endpoint precedes the right endpoint located at the same point of the real line. Then move from left to right counting the number of layers. When we cross the left endpoint, increase the number of layers by 1; when we cross the right endpoint, decrease the number of layers by 1. Please note that two adjacent intervals are processed correctly; that is, the left endpoint precedes the right endpoint according to our convention.] □

**4.3.3.** Assume that n points in the plane are given. Find a polygonal arc with n − 1 sides whose vertices are the given points, and whose sides do not intersect. (Adjacent sides may form a 180° angle.) The number of operations should be of order n log n.

*Solution.* Sort all the points with respect to the $x$-coordinate; when $x$-coordinates are equal, take the $y$-coordinate into account, then connect all vertices by line segments (in that order). □

**4.3.4.** The same problem for a polygon: for a given set of points in the plane find a polygon having these points as vertices.

*Solution.* Take the leftmost point (the point whose $x$-coordinate is minimal). Consider all the rays starting from this point and going through all other points. Sort these rays according to their slopes, and sort the points that are on the same ray according to their distance from the initial point. The polygon goes from the initial point along the ray with minimal slope, then visits all the points in the order chosen, returning via the ray with maximal slope (where points are visited in the reversed order). □

**4.3.5.** Assume that n points in the plane are given. Find their *convex hull*; that is, the minimal convex polygon that contains all the points. (A rubber band put on the nails is the convex hull of the nails inside it.) The number of operations should be of order n log n.

[*Hint.* Order all the points according to one of the orderings mentioned in the two preceding problems. Then construct the convex hull considering the points one by one. (To maintain information about the current convex hull, it is useful to use a deque; see chapter 6, page 91. It is not necessary, however, when the points are ordered according to their slopes.)]                                    □

## 4.4 Lower bound for the number of comparisons

Suppose we have $n$ objects (say, stones) of different weights and a balance that can be used to find which of any two given stones is heavier. In programming terms, we have access to a Boolean function `heavier(i,j:1..n)`. Our goal is to sort all the stones in increasing order using the balance as few times as possible (making the fewest calls to the function `heavier`).

Of course, the number of comparisons depends not only on the algorithm we choose but also on the initial order of the stones. By *complexity* of the algorithm we mean the number of comparisons *in the worst case*.

**4.4.1.** Prove that any sorting algorithm for $n$ stones has complexity at least $\log_2 n!$. (Here $n! = 1 \cdot 2 \cdots n$.)

*Solution.* Assume that we have an algorithm of complexity $d$; that is, an algorithm that makes at most $d$ comparisons (in all cases). For any of $n!$ possible orderings of the stones let us write down the results of all the comparisons (values returned by calls to the function `heavier`). The protocol is a binary string of length at most $d$. If necessary, pad it with trailing zeros to get a string of length $d$. Now we have $n!$ binary strings of length $d$ (corresponding to $n!$ permutations of input stones). All those strings are different, otherwise our algorithm gives the same answer for two different orderings (and at least one of the answers is incorrect). Therefore, $2^d \geqslant n!$.

Another way to say the same thing is to consider the tree of possibilities that appear during the execution of the algorithm. Indeed, a tree of height $d$ has no more than $2^d$ leaves.

This argument shows that any algorithm that relies upon comparisons and exchange operations only, requires at least $\log_2 n!$ comparisons. A simple calculation shows that $\log_2 n! \geqslant \log_2(n/2)^{n/2}$ (we omit the first half of the factors and replace the remaining factors by $n/2$). Now $\log_2(n/2)^{n/2} = (n/2)(\log_2 n - 1) \geqslant Cn \log_2 n$ for some $C$. Therefore, our sorting algorithms are close to optimal (improvement is limited to a constant factor).                                    □

However, a sorting algorithm that uses not only comparisons (but also the internal structure of the sorted objects) may be faster. Here is an example:

**4.4.2.** An integer array a[1]..a[n] is given; all the integers are nonnegative and do not exceed m. Sort this array using no more than $C(m+n)$ operations ($C$ is a constant that does not depend on m and n).

*Solution.* For each number in 0..m, count how many times it appears in the array. (These data can be collected during one pass through the array.) Then erase the array and write down its elements in increasing order using the information about the multiplicity of each number. □

Note that this algorithm does not exchange elements of the array but puts "fresh" sorted numbers into the array.

There exists another sorting method that sequentially performs several "partial sorts" with respect to fixed bits. Let us start with the following problem:

**4.4.3.** Rearrange the array a[1]..a[n] in such a way that all even elements precede all odd elements (not changing the order inside each of the two groups).

*Solution.* Copy all the even elements into an auxiliary array. Then append all the odd elements to the auxiliary array and finally copy all elements back. □

**4.4.4.** An array of $n$ integers in the range $0, \ldots, 2^k - 1$ is given. Each integer is written as a binary string of length $k$. Using the tests "$i$-th bit is 0" and "$i$-th bit is 1" instead of comparisons, sort all the integers. The number of operations should be of order $nk$.

*Solution.* Sort all the numbers with respect to the last bit as in the preceding problem. Then sort them with respect to the bit next the last one, etc. After $k$ stages, the numbers will be sorted. Indeed, by induction over $i$, we can easily prove the following statement: "after $i$ steps, any two numbers that differ only in the last $i$ bits, are in the correct order". (Or prove by induction the following statement: "after $i$ steps the suffixes of length $i$ are in the right order".) □

A similar algorithm can be constructed using $m$-ary notation instead of binary. The following problem is useful in this regard.

**4.4.5.** Assume that an array of $n$ elements and a function $f$ defined on those elements are given. Assume that the possible values of $f$ are $1, \ldots, m$. Rearrange the array in such a way that the values of $f$ are in non-decreasing order and the elements with equal values of $f$ are in the same order as before. The number of operations should be of order $m + n$.

[*Hint.* Create $m$ lists of total length $n$ using "pointer implementation" (see chapter 6, p. 87). Put an element into the $i$-th list if the value of $f$ is equal to $i$. Another possibility: count how many elements have a given value of $f$ (for all $m$ possible values); thereafter, we know where the elements of any given $f$-value should be placed in the array.] □

## 4.5 Problems related to sorting

**4.5.1.** Find the minimal complexity (= the number of comparisons in the worst case) for an algorithm that finds the stone with minimal weight.

*Solution.* The obvious algorithm with the invariant relation "the minimal among the first $i$ stones is found" requires $n-1$ comparisons. No algorithm can have smaller complexity. This is a corollary of a stronger statement, see the next problem.    □

**4.5.2.** An expert wants to convince a jury that a given stone has minimal weight among $n$ given stones. The expert wants to do this using a balance less than $n-1$ times. Prove that this is impossible. (The expert knows in advance the weights of all the stones; the jury does not.)

*Solution.* Consider stones as vertices of a graph. For any comparison, draw an edge between the corresponding pair of vertices. After $n-1$ measurements, the graph is not connected; it has more than 1 connected component, because each edge decreases the number of connected components by at most 1. Therefore, the jury knows nothing about the relation between weights of stones from different components and may assume that the stone with minimal weight is in any of the components.    □

Let us stress the difference between this problem and the preceding one. In this problem, we have to show that $n-2$ comparisons are not enough to prove that a given stone has minimal weight even if we know the answer in advance — not to mention finding the answer. (The difference between the two settings is clear in the case of sorting. When a correct answer is known, it can be confirmed by $n-1$ comparisons (each stone should be compared with the next one), which is many fewer comparisons than what was needed to find the answer.)

**4.5.3.** Prove that it is possible to find the stones with minimal and maximal weights among $2n$ stones using only $3n-2$ comparisons.

*Solution.* Let us group $2n$ stones into $n$ pairs and compare stones within each pair. We have $n$ "winners" and $n$ "losers". Then we need $n-1$ comparisons to find the winner among the winners and $n-1$ comparisons to find the loser among the losers.    □

**4.5.4.** Prove that no algorithm can find the stones with minimal and maximal weights among $2n$ stones using less than $3n-2$ comparisons in the worst case.

*Solution.* Assume that such an algorithm exists. When it is applied to a group of $2n$ stones, we observe the changes in four quantities: the numbers of stones that

(a) have lost at least one game and have won at least one game;
(b) have lost at least one game but never won a game;
(c) have won at least one game but never lost a game;
(d) never lost a game and never won a game (i.e., never played)

An $a$-stone could be neither the (total) winner nor the loser. A $b$-stone could be a loser but not a winner; a $c$-stone could be a winner but not a loser. Finally, a $d$-stone still has a chance to be either a winner or a loser. Let us denote by $a, b, c, d$ the numbers of stones of all four types. The following table shows the possible changes is the parameters after a comparison between two stones of some types is made:

| comparison | $a$ | $b$ | $c$ | $d$ | $b + c + (3/2)d$ |
|---|---|---|---|---|---|
| $a$–$a$ | 0 | 0 | 0 | 0 | 0 |
| $a > b$ | 0 | 0 | 0 | 0 | 0 |
| $a < b$ | +1 | −1 | 0 | 0 | −1 |
| $a < c$ | 0 | 0 | 0 | 0 | 0 |
| $a > c$ | +1 | 0 | −1 | 0 | −1 |
| $a > d$ | 0 | +1 | 0 | −1 | −1/2 |
| $a < d$ | 0 | 0 | +1 | −1 | −1/2 |
| $b$–$b$ | +1 | −1 | 0 | 0 | −1 |
| $b < c$ | 0 | 0 | 0 | 0 | 0 |
| $b > c$ | +2 | −1 | −1 | 0 | −2 |
| $b < d$ | 0 | 0 | +1 | −1 | −1/2 |
| $b > d$ | +1 | 0 | 0 | −1 | −3/2 |
| $c$–$c$ | +1 | 0 | −1 | 0 | −1 |
| $c < d$ | +1 | 0 | 0 | −1 | −3/2 |
| $c > d$ | 0 | +1 | 0 | −1 | −1/2 |
| $d$–$d$ | 0 | +1 | +1 | −2 | −1 |

The last column shows how the weighted sum $s = b + c + (3/2)d$ changes. (Intuitively, $s$ measures the amount of remaining work: a stone for which there is no information at all, is 1.5 times more difficult than a stone for which one-sided information is available.)

Initially $s = 3n$; when algorithm terminates, $s = 2$ (all the stones, except the winner and the loser, are $a$-stones; the winner is a $c$-stone and the loser is a $b$-stone). The table shows that every comparison has one "unlucky" outcome when $s$ decreases by 1 (or even less). These unlucky outcomes are real (i.e., do not contradict the results of previous comparisons). Indeed, when a $b$-stone is compared with a $c$-stone it is possible that the $c$-stone wins (there is no upper bound for its weight); when a $d$-stone is compared with any other stone, the result can be arbitrary (there is no restriction for the weight of a $d$-stone). (Alternatively, we may note that if one of the outcomes of a comparison is impossible, this comparison can be omitted.)

If all the comparisons have unlucky outcomes, we need ad least $3n - 2$ comparisons to go down from $3n$ to 2. $\square$

**4.5.5.** Assume that $n$ stones of different weights are given. Find both the stone with maximal weight and the second best using at most $n + \lceil \log_2 n \rceil - 2$ comparisons. (Here $\lceil \log_2 n \rceil$ is the minimal integer $k$ such that $2^k \geqslant n$.)

*Solution.* First we find the winner and then the second best. It is clear that the only possible candidates for the second best element are those who lost the game to the winner (if a stone $x$ is lighter than some other stone except the winner, then $x$ cannot be the second best).

Let us use an Olympic tournament to find the winner (stones are grouped into pairs, in each pair the loser is discarded, the winners are again grouped into pairs etc.). Then we need $k$ rounds for $2^k$ stones and $\lceil \log_2 n \rceil$ rounds for $n$ stones. After each comparison one stone is discarded, so we need $n - 1$ comparisons to find the winner and then $\lceil \log_2 n \rceil - 1$ comparisons to find the second best among the candidates beaten by the winner.                                                                □

**4.5.6.** Prove that no algorithm can find both the winner and the second best stone (among $n$ stones of different weights) using less that $n + \lceil \log_2 n \rceil$ comparisons in the worst case.

*Solution.* Assume the some algorithm is applied to a group of stones. At any moment by $k_i$ we denote the number of stones that have lost exactly $i$ games (comparisons). We count only direct comparisons; if two comparisons give $a < b$ and $b < c$, we say that $a$ has lost one game (to $b$), not two.

Evidently, the sum of all $k_i$ equals the number of games (each game has one loser). Therefore it is enough to show that (for any algorithm) it could happen that $k_1 + k_2 \geqslant n + \lceil \log_2 n \rceil - 2$. Let us show the outcomes that guarantee this inequality.

By "leaders" we mean stones that have not lost a game yet. Initially we have $n$ leaders; at the end there is only one (since every leader is a potential winner). Therefore $k_1 \geqslant n - 1$ (all the players except the winner have lost at least one game).

Let us choose the outcomes as follows. When two non-leaders meet, any outcome is OK. When a leader meets a non-leader, the leader wins. When two leaders meet, a more experienced one wins (experience is the number of games the leader won); ties are broken arbitrarily.

This guarantees that $k_2 \geqslant \lceil \log_2 n \rceil - 1$ (and this gives the required inequality for $k_1 + k_2$). To prove this, let us introduce the subordination relation: each player is attached to one of the leaders. Initially every player is a leader and is attached to itself. When a leader meets a non-leader (or two non-leaders meet), attachment relation does not change. When two leaders meet, the loser and all players attached to it become attached to the winner.

A simple induction shows that leader with experience $k$ has at most $2^k$ players attached to it. Initially $k = 0$ and this group is a singleton. If a leader with experience $k$ wins a game against some other leader with experience at most $k$, each of them has at most $2^k$ attached players which gives at most $2^{k+1}$ players attached to a player with experience $k + 1$.

Therefore, at the end the winner has experience at least $\lceil \log_2 n \rceil$ (all players are attached to it). All its partners (except the second best) have lost one more game (otherwise there still could be the second best candidates), which gives the required bound for $k_2$.                                                                □

**4.5.7.** Prove that the same bound is still valid if we need to find only the second best player (and are not interested in the best one).

[*Hint.* When the second best player is declared, this player has lost at least one game (otherwise it could be a winner), and the winner of this game is the best.]                                                                □

**4.5.8.** Assume that $n$ stones of different weights are given. Let $k$ be a number in the range $1, \ldots, k$. Find the $k$-th stone (in the order of increasing weights) making not more than $Cn$ comparisons, where $C$ is some constant that does not depend on $k$ or $n$.

*Remark.* Using sorting, we can do this in $Cn \log n$ steps. See chapter 7, p. 118, where a hint for this (rather difficult) problem is given. □

The following problem has a surprisingly simple solution.

**4.5.9.** There are $n$ stones that look identical, but in fact, some of them have different weights. There is a device that can be applied to two stones and tells whether they are different or not (but it does not say which one is heavier). It is known in advance that most of the stones (more than 50%) are identical. Find one of those identical stones making no more than $n$ comparisons. (Beware: it is possible that two stones are identical but do not belong to the majority of identical stones.)

[*Hint.* If two stones are different, they may be both discarded, because one of them does not belong to the majority and the majority survives.]

*Solution.* The program processes the stones one-by-one and keeps the number of the stones processed in a variable i. (Assume that stones are numbered 1..n). The program remembers the number of the "current candidate" c and its "multiplicity" k. The names are explained by the following invariant relation (I):

> If we add k copies of the c-th stone to the unprocessed stones (i+1..n), the majority stones in the initial array will remain the majority in the new array.

Here is the program:

```
k:=0; i:=0;
{(I)}
while i<>n do begin
  if k=0 then begin
    k:=1; c:=i+1; i:=i+1;
  end else if (i+1-th stone is the same as c-th)
       then begin
    i:=i+1; k:=k+1;
    {replace a physical stone by a virtual stone}
  end else begin
    i:=i+1; k:=k-1;
    {discard one physical and one virtual stone}
  end;
end;
c-th stone is the answer
```

*Remark.* All three branches of the if-block include the statement i:=i+1, so it can be moved to the upper level. □

Let us mention that this program finds the most frequent stone only if it forms the majority (more than 50%).

❦ This problem can be found as problem 4-7 on page 75 of the book [3] in a completely different setting ("VLSI chip testing") where a recursive solution is sketched.

At first glance, the following problem seems unrelated to sorting.

**4.5.10.** There is a square array $a[1..n, 1..n]$ filled by 0s and 1s. It is known that for some $i$ the $i$-th row contains only 0s and at the same time the $i$-th column contains only 1s (except the main diagonal entry $a[i,i]$, which may be arbitrary). Find this $i$ (which is unique). The number of operations should be of order $n$. (Please note that the number of operations should be much smaller than the total number of elements in $a$.)

[*Hint*. Assume we get the Boolean value $a[i][j]$ when comparing two virtual stones with numbers $i$ and $j$. Recall that the maximal element among $n$ elements can be found using $n-1$ comparisons. Take into account that the array may not be "transitive"; however, after two numbers are compared, one of them may be discarded.] □

# 5

# Finite-state algorithms in text processing

This chapter describes a simple technique often used to process input, change text encodings etc. We consider two examples of "lexical analysis" that can occur during the first pass of a compiler. In section 5.1 we show how to process multi-character symbols. Then in section 5.2 we show how a finite automaton can convert string representation of a number to its numeric value. More advanced applications of finite-state machines are described in chapter 10.

## 5.1 Compound symbols, comments, etc.

**5.1.1.** Throughout a program text the operation $x^y$ was denoted by x**y. It was decided that notation should be changed to x^y. How do we do that? The input text is read character-by-character; the output text should be produced in the same manner.

*Solution.* At any time, the program is in one of two states: "basic" state and "after" state (after an asterisk):

| State | Next symbol | New state | Action |
|-------|-------------|-----------|--------|
| basic | * | after | none |
| basic | $x \neq *$ | basic | print $x$ |
| after | * | basic | print ^ |
| after | $x \neq *$ | basic | print *, $x$ |

If after reading all the text, the program is in the "after" state, it should print an asterisk (and quit). □

*Remark.* Our program replaces *** by ^* (and not by *^). We did not specify the behavior of the program in this case, assuming (as is often done) that some "reasonable" behavior is expected. In this example, the simplest way to describe the required behavior is to list the states and the corresponding actions.

Please note also that if two asterisks appear in other parts of the program (say, comments), they will be also replaced.

A. Shen, *Algorithms and Programming*, Springer Undergraduate Texts in Mathematics and Technology, DOI 10.1007/978-1-4419-1748-5_5, © Springer Science+Business Media, LLC 2010

**5.1.2.** Write a program that deletes all occurrences of the substring abc.    □

**5.1.3.** In Pascal, comments are surrounded by curly braces like this:

```
begin {here a block begins}
i:=i+1; {increase i by one}
```

Write a program that removes all comments and puts a space character in the place of a removed comment. (According to Pascal rules, 1{one}2 is equivalent to 1 2, not 12).

*Solution.* The program has two states: a "basic" state and an "inside" state (inside a comment).

| State | Next symbol | New state | Action |
|-------|-------------|-----------|--------|
| basic | { | inside | none |
| basic | $x \neq ($ | basic | print $x$ |
| inside | } | basic | print a space |
| inside | $x \neq \}$ | inside | none |

□

This program cannot deal with nested comments: the string

```
{{comment inside a} comment}
```

is transformed into

```
 comment}
```

(the latter string starts with two spaces). It is impossible to deal with nested comments using a finite automaton (a program that has finite number of internal states); roughly speaking, we have to remember the number of opening braces and a finite automaton cannot do that.

Please note that after reading all the text, the program may still be in the "inside" state. Most probably, we would like to consider this as an error.

**5.1.4.** Pascal programs also contain quoted strings. If a curly brace appears inside a string, it does not mean the start of a comment. Similarly, a quote symbol inside a comment does not signify a string. How do we modify the above program to take this into account?

[*Hint.* We need three states: "basic", "inside a comment", "inside a string".]    □

(Note that actual Pascal conventions are more complicated allowing a quote to appear inside a quoted string, etc.)

**5.1.5.** One more feature that exists in many Pascal implementations is a comment of the type

```
i := i+1;   (* here i is increased by 1 *)
```

A closing comment symbol must be paired with an opening comment symbol of the same type (e.g., {...*) is not permitted). How do we deal with these types of comments?    □

## 5.2 Numbers input

Assume that a program scans a decimal representation of some number from left to right. The program should "read" this number; that is, put its value into a variable of type `real`. Also, the program should complain if the input is incorrect.

Let us specify the problem in more detail. Assume that the input string is divided into two parts: the part that is already processed and the remaining part. We have access to a function Next:char, which returns the first symbol of the unprocessed part. Also, we have access to a procedure Move, which moves the first unprocessed symbol to the processed part.

| processed part | Next | ? | ? |
|---|---|---|---|

By a decimal number we mean a character string of the type

⟨0 or more spaces⟩ ⟨1 or more digits⟩

or

⟨0 or more spaces⟩ ⟨1 or more digits⟩ . ⟨1 or more digits⟩

Please note that this definition does not allow the following strings:

1.        .1        1.␣1        −1.1

Let us now state the problem:

**5.2.1.** Read the maximal prefix of the input string that may be a prefix of a decimal number. Determine whether this prefix is a decimal number or not.

*Solution.* Let us write a program using Pascal's "enumeration type" for clarity. (The variable state may have one of the listed values.)

```
var state:
  (Accept, Error, Initial, IntPart, DecPoint, FracPart);

state := Initial;
while (state <> Accept) or (state <> Error) do begin
  if state = Initial then begin
    if Next = ' ' then begin
      state := Initial; Move;
    end else if Digit(Next) then begin
      state := IntPart;
        {after the start of the integer part}
      Move;
```

```
      end else begin
      | state := Error;
      end;
   end else if state = IntPart then begin
      if Digit (Next) then begin
      | state := IntPart; Move;
      end else if Next = '.' then begin
      | state := DecPoint; {after the decimal point}
      | Move;
      end else begin
      | state := Accept;
      end;
   end else if state = DecPoint then begin
      if Digit (Next) then begin
      | state := FracPart; Move;
      end else begin
      | state := Error; {at least one digit is needed}
      end;
   end else if state = FracPart then begin
      if Digit (Next) then begin
      | state := FracPart; Move;
      end else begin
      | state := Accept;
      end;
   end else if
   | {this cannot happen}
   end;
end;
```

Please note that the assignments state := Accept and state := Error are not accompanied by a call to procedure Move, so the symbol after the end of the decimal number is left unprocessed.                                                                    □

This program does not store the value of the number.

**5.2.2.** Add the following requirement to the preceding program: If a processed part is a decimal number, its value should be placed into the variable val : real.

*Solution.* While reading the fractional part, we use the variable scale which is a factor for the digit to come (0.1, 0.01 etc.).

```
state := Initial; val:= 0;
while (state <> Accept) or (state <> Error) do begin
   if state = Initial then begin
      if Next = ' ' then begin
      | state := Initial; Move;
      end else if Digit(Next) then begin
```

```
            state := IntPart;
              {after the start of the integer part}
            val := DigitVal(Next);
            Move;
          end else begin
            state := Error;
          end;
        end else if state = IntPart then begin
          if Digit (Next) then begin
            state := IntPart; val := 10*val + DigitVal(Next);
            Move;
          end else if Next = '.' then begin
            state := DecPoint; {after the decimal point}
            scale := 0.1;
            Move;
          end else begin
            state := Accept;
          end;
        end else if state = DecPoint then begin
          if Digit (Next) then begin
            state := FracPart;
            val := val + DigitVal(Next)*scale;
            scale := scale/10;
            Move;
          end else begin
            state := Error; {at least one digit is needed}
          end;
        end else if state = FracPart then begin
          if Digit (Next) then begin
            state := FracPart;
            val := val + DigitVal(Next)*scale;
            scale := scale/10;
            Move;
          end else begin
            state := Accept;
          end;
        end else if
          {this cannot happen}
        end;
      end;
    end;                                                       □
```

**5.2.3.** Repeat the previous problem if the number may be optionally preceded by − or +. □

The format of numbers in this problem can be represented as follows:

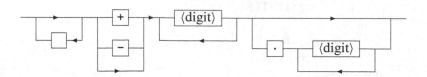

**5.2.4.** The same problem if the number may be followed by an integer exponent, as in 254E–4 (= 0.0254) or 0.123E+9 (= 123 000 000). Draw the corresponding picture.                                                                                    □

**5.2.5.** What changes in the above program above are necessary to allow empty integer or fractional parts like in 1., .1 or even . (the latter number is considered to be equal to zero)?                                                                          □

We return to finite-state algorithms (also called *finite automata*) in chapter 10.

# 6

# Data types

It is convenient to describe algorithms using appropriate data types. Basically, a data type is a set of values (permissible for the variables of this type) and a list of allowed operations. Data types are important since they separate two aspects: (1) what kind of information we want to keep and what we want to do with it, and (2) how this information is represented in our program, and, ultimately, in the computer's memory.

In this chapter we consider some basic data types (stacks, queues, sets, priority queues) and show how they can be implemented and used.

## 6.1 Stacks

Let T be some type. Consider the data type "stack of elements of type T." Values of that type are sequences of values of type T.

Operations:

- Make_empty (var s: stack of elements of type T)
- Add (t:T; var s: stack of elements of type T)
- Take (var t:T; var s: stack of elements of type T)
- Is_empty (s: stack of elements of type T): Boolean
- Top (s: stack of elements of type T): T

(We use Pascal notation even though the stack type does not exist in Pascal.) The procedure "Make_empty" makes the stack empty. The procedure "Add" adds t to the end of the sequence s (i.e., the top of the stack). The procedure "Take" is applicable if the sequence s is not empty; it takes the last element away from s and puts it into the variable t. The expression "Is_empty(s)" is true when the sequence s is empty. The expression "Top(s)" is defined when s is not empty; its value is the last element of the sequence s.

Usually the operations "Add" and "Take" are called "Push" and "Pop" respectively; we use the names "Add" and "Take" to stress the similarity between stacks and queues (section 6.2).

A. Shen, *Algorithms and Programming*, Springer Undergraduate Texts in Mathematics and Technology, DOI 10.1007/978-1-4419-1748-5_6, © Springer Science+Business Media, LLC 2010

Our goal is to show how stacks can be implemented in Pascal and what they can be used for.

**Stack: array implementation**

Assume that the number of elements in a stack never exceeds some constant n. Then the stack can be implemented using two variables:

```
Content: array [1..n] of T;
Length: integer;
```

We assume that our stack contains elements

```
Content [1],...,Content [Length]
```

- To make the stack empty, it is enough to perform the assignment

```
Length := 0
```

- Adding element t:

```
{Length < n}
Length := Length + 1;
Content [Length] :=t;
```

- Taking element into a variable t:

```
{Length > 0}
t := Content [Length];
Length := Length - 1;
```

- The stack is empty when Length = 0.
- The top of the stack is Content [Length], assuming Length > 0.

Therefore, a variable of type stack can be replaced in a Pascal program by two variables Content and Length. We can also define the type stack as follows:

```
const n = ...
type
  stack = record
    Content: array [1..n] of T;
    Length: integer;
  end;
```

We then define procedures dealing with stack variables. For example, we write

```
procedure Add (t: T; var s: stack);
begin
  {s.Length < n}
  s.Length := s.Length + 1;
  s.Content [s.Length] := t;
end;
```

**The use of stacks**

In the following problem, we consider sequences of opening and closing parentheses ( ) and square brackets [ ]. Some sequences are considered to be "correct". Namely, a sequence is correct if its correctness follows from the following rules:

- the empty sequence is correct;
- if $A$ and $B$ are correct, then $AB$ is correct;
- if $A$ is correct, then $[A]$ and $(A)$ are correct.

Example. The sequences (), [[]], [()[]()][] are correct, while the sequences ],)(, (], ([)] are not.

**6.1.1.** Check the correctness of a given sequence. The number of operations should be proportional to the length of the sequence. We assume that the sequence terms are encoded as follows:

$$( \quad 1$$
$$[ \quad 2$$
$$) \ -1$$
$$] \ -2$$

*Solution.* Let a[1]..a[n] be a sequence of length n. Consider a stack whose elements are opening parentheses and brackets (i.e., the numbers 1 and 2).

Initially the stack is empty. We scan the sequence from left to right. When an opening parenthesis or bracket is found, we put it onto the stack. When a closing parenthesis or bracket is found, we check if the top of the stack is a complementary parenthesis or bracket. If not, we stop and reject the input. If so, we take the top of the stack away. The sequence is correct if it is not rejected while reading the input and if the stack is empty after the input is exhausted.

```
Make_empty (s);
i := 0; Error_found := false;
{i symbols are processed}
while (i < n) and not Error_found do begin
  i := i + 1;
  if (a[i] = 1) or (a[i] = 2) then begin
    Add (a[i], s);
  end else begin   {a[i] is either -1 or -2}
    if Is_empty (s) then begin
      Error_found := true;
    end else begin
      Take (t, s);
      Error_found := (t <> - a[i]);
    end;
  end;
end;
Correct := (not Error_found) and Is_empty (s);
```

Let us prove the correctness of our program.

(1) If the input sequence is correct, our program accepts it. This can be proved by induction. We need to prove that (a) our program accepts the empty sequence; (b) that it accepts the sequence $AB$ (assuming that $A$ and $B$ are accepted); and (c) it accepts the sequences $[A]$ and $(A)$ assuming that $A$ is accepted.

An empty sequence is accepted for obvious reasons. (Note: In this case, the while-loop is not executed.)

For $AB$ our program works exactly as for $A$ until all symbols of $A$ are processed; therefore, the stack is empty at that moment. Then program processes $B$ (and finishes with the empty stack, because $B$ is accepted by assumption).

For $[A]$ the program begins by putting an opening bracket onto the stack. Then the program processes $A$, the only difference is that there is an additional bracket at the bottom of the stack, and it never interferes with the program's execution. When $A$ is finished, the stack is empty except for the left bracket; at the next step, the stack becomes empty. A similar thing happens for $(A)$.

(2) Let us now prove that if the program accepts some sequence, then the sequence is correct. This is proved by induction over the length of the sequence. Consider the length of the stack during execution. If the stack becomes empty at some point, then the sequence can be divided into two parts and each of the parts is accepted by the program. Therefore, each part is correct (inductive hypothesis) and the sequence as a whole is correct (definition of correctness). Now assume that the stack never becomes empty (except for the beginning and the end). This means that the bracket or parenthesis put onto the stack at the first step is removed at the last step. Therefore, the first and last symbols in our sequence are complementary, the sequence is of type $(A)$ or $[A]$, and the behavior of the program differs from its behavior on $A$ only by the additional parenthesis or bracket at the bottom of the stack. Therefore, by the induction hypothesis, $A$ is correct and the sequence is correct by definition.                                                                                    □

**6.1.2.** The program can be simplified if the sequence contains only parentheses and no brackets. How?

*Solution.* In this case, the stack is reduced to its length, and we arrive at the following statement: A sequence of "(" and ")" is correct if and only if each prefix contains no more symbols ")" than "(", and the entire sequence has equal numbers of both symbols.                                                                                    □

**6.1.3.** Implement two stacks using one array. The total number of elements in both stacks is limited by the array length; all stack operations should run in $O(1)$ time (i.e., running time should be bounded by a constant).

*Solution.* The stacks grow in opposite directions starting from the beginning and end of the array Content[1..n]. One stack occupies places

$$Content[1]..Content[Length1],$$

while the other stack occupies places

Content[n]..Content[n-Length2+1]

(both stacks are listed from bottom to top). Stacks do not overlap if their total length does not exceed n.                                                                □

**6.1.4.** Implement k stacks of elements of type T with a total of at most n elements using arrays with total length $C(n + k)$. Each stack operation (except initialization, which makes all stacks empty) should be performed in constant time (not depending on n and k). (In other words, the implementation should require space $O(n + k)$ and run in time $O(1)$ for each operation.)

*Solution.* We use a "pointer implementation" of stacks. It uses three arrays:

```
Content: array [1..n] of T;
Next: array [1..n] of 0..n;
Top: array [1..k] of 0..n;
```

The array Content can be thought of as n cells numbered from 1 to n. Each of the cells is capable of holding one element of type T. The array Next is represented by arrows between elements: there is an arrow from i to j if Next[i]=j. (If Next[i]=0, there are no arrows from i.) The content of the s-th stack (s ∈ {1..k}) is determined as follows: the top element is Content[Top[s]] and other elements are read by following the arrow links (if they exist). Moreover,

$$(\text{s-th stack is empty}) \Leftrightarrow \texttt{Top[s]} = 0.$$

The "arrow trajectories" starting from

$$\texttt{Top[1]}, \ldots, \texttt{Top[k]}$$

(those not equal to 0) are disjoint. Besides these, we need one more trajectory that traverses all locations that are currently free. Its starting point is stored in the variable Free: 0..n (where Free = 0 means that all places are occupied). Here is an example:

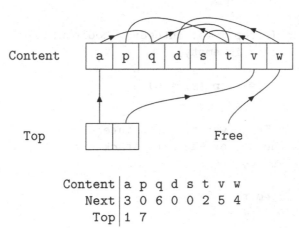

```
Content a p q d s t v w
   Next 3 0 6 0 0 2 5 4
    Top 1 7
```

$$Free = 8$$

Stacks: the first one contains p, t, q, a (a is on the top); the second one contains s, v (v is on the top).

```
procedure Initialize; {Make all stacks empty}
| var i: integer;
begin
| for i := 1 to k do begin
| | Top [i]:=0;
| end;
| for i := 1 to n-1 do begin
| | Next [i] := i+1;
| end;
| Next [n] := 0;
| Free:=1;
end;

function Is_free: Boolean;
begin
| Is_free := (Free <> 0);
end;

procedure Add (t: T; s: integer);
| {Add t to the s-th stack}
| var i: 1..n;
begin
| {Is_free}
| i := Free;
| Free := Next [i];
| Next [i] := Top [s];
| Top [s] :=i;
| Content [i] := t;
end;

function Is_empty (s: integer): Boolean;
| {s-th stack is empty}
begin
| Is_empty := (Top [s] = 0);
end;

procedure Take (var t: T; s: integer);
| {Take the top of the s-th stack into t}
| var i: 1..n;
| begin
| {not Is_empty (s)}
| i := Top [s];
```

```
  t := Content [i];
  Top [s] := Next [i];
  Next [i] := Free;
  Free := i;
end;

function Top_element (s: integer): T;
  {Top of the s-th stack}
begin
  Top_element := Content[Top[s]];
end;
```
                                                                            □

## 6.2 Queues

Values of type "queue of elements of type T" are sequences of values of type T. The same is true for stacks, but the difference is that queue elements are added to the beginning of a sequence and are taken from the end of it. Therefore, an element that arrived first to a queue will be the first element taken from it. Hence the name First In First Out (FIFO), which is used for queues. The rule used for stacks is called Last In First Out (LIFO).

Operations on queues:

- Make_empty (var x: queue of elements of type T);
- Add (t:T, var x: queue of elements of type T);
- Take (var t:T, var x: queue of elements of type T);
- Is_empty (x: queue of elements of type T): Boolean;
- First_element (x: queue of elements of type T): T.

The procedure "Add" adds the specified element to the end of the queue. The procedure "Take" is applicable if the queue is not empty; it puts the first element of the queue into a variable t, removing it from the queue. (The first element is the longest-waiting element.)

The procedures "Add" and "Take" are often called "Enqueue" and "Dequeue".

**Queue: array implementation**

**6.2.1.** Implement a queue of limited size in such a way that all operations run in $O(1)$ time (that is, in time not exceeding some constant, which does not depend on length of the queue).

*Solution.* Assume that queue elements are stored as consecutive elements in an array. The queue grows to the right and is taken from the left. A growing queue may reach the end of the array, so we assume the array is "wrapped around" in circular fashion.

Our implementation uses an array

```
Content: array [0..n-1] of T
```

and variables

```
First: 0..n-1
Length : 0..n
```

The queue is formed by elements

```
Content [First], Content [First + 1],..., Content [First+Length-1]
```

where addition is performed modulo n. (Warning: If you instead use variables `First` and `Last` whose values are residues modulo n, be careful not to mix the empty queue with the queue containing n elements.)

The queue operations are implemented as follows:

Make_empty:

```
Length := 0;
First := 0;
```

Add an element t:

```
{Length < n}
Content [(First + Length) mod n] := t;
Length := Length + 1;
```

Take element into variable t:

```
{Length > 0}
t := Content [First];
First := (First + 1) mod n;
Length := Length - 1;
```

Is_empty:

```
Length = 0
```

First_element:

```
Content [First]
```
□

**6.2.2.** (Communicated by A.G. Kushnirenko) Implement a queue using two stacks (and a fixed number of variables of type T). For *n* queue operations starting with an empty queue, the implementation should perform not more than *Cn* stack operations.

*Solution.* We maintain the following invariant relation: *stacks whose bottoms are put together, form the queue.* (In other words, listing all elements of one stack from top to bottom and then of the other stack from bottom to top, we list all the queue elements in the proper order.) To add an element to the queue, it is enough to add it to one of the stacks. To check if the queue is empty, we must check that both stacks are empty. When taking the first element from the queue, we should distinguish between two cases. If the stack that contains the first element is not empty, there is no problem. If that stack is empty, the required element is buried under all the elements of the second stack. In this case, we move all the elements one-by-one onto the first stack (their ordering is reversed) and return to the first case.

The number of operations for this step is not limited by any constant. However, the requirement posed in the problem is still met. Indeed, any element of the queue can participate in such a process at most once during its presence in the queue. □

**6.2.3.** Deque (double-ended queue) is a structure that combines the properties of a queue and a stack: we can add and remove elements from both ends of a deque. Implement a deque using an array in such a way that each deque operation runs in $O(1)$ time. □

**6.2.4.** (Communicated by A.G. Kushnirenko.) A deque of elements of type T is given. The deque contains several elements. The program should determine how many elements are in the deque. Program may use variables of type T and integer variables, but arrows are not allowed.

[*Hint.* (1) We can perform a cyclic shift on deque elements taking an element from one end and adding it to the other end. After $n$ shifts in one direction, we return the deque to its initial state by $n$ shifts in the other direction. (2) How do we know that the cycle is complete? If we know in advance that some element is guaranteed not to appear in the deque, this is easy. We put this "signal" element into the deque and wait until it appears at the other end. But we do not have such an element. Instead, we may perform (for any fixed $n$) a cyclic shift by $n$ positions twice adding two different elements. If the elements that appear after the shift are different, we have made a complete cycle.] □

**Queue applications**

**6.2.5.** (see E.W. Dijkstra's book [5]) Print in increasing order the first n positive integers whose factorization contains only the factors 2, 3, and 5.

*Solution.* The program uses three queues x2, x3, x5. They are used to store elements which are 2, 3, and 5 times larger than already printed elements, but are not yet printed. We use the procedure

```
procedure Print_and_add (t: integer);
begin
  writeln (t);
  Add (2*t, x2);
  Add (3*t, x3);
  Add (5*t, x5);
end;
```

The program is as follows:

```
.. make queues x2, x3, x5 empty
Print_and_add (1);
k := 1;
{invariant relation: k first elements of the required set
    are printed; the queues contain (in increasing order)
    elements that are 2, 3 and 5 times bigger than the
    elements already printed, but are not printed yet}
while k <> n do begin
    x := min (Next(x2), Next(x3), Next(x5));
    Print_and_add (x);
    k := k+1;
    .. take x from the queues where it was present;
end;
```

Let us check the correctness of the program. Assume that the invariant relation is valid and we perform the operations as prescribed. Let x be the smallest element of our set that is not printed. Then x is larger than 1, and it is divisible by 2, 3, or 5. The quotient belongs to the set, too. The quotient is smaller than x and is therefore printed. Thus x is present in one of the queues. It is the smallest element in any queue to which x belongs (because all the elements less than x are already printed and cannot appear in any queue). When x is printed, we must delete x from the queues and add the corresponding multiples of x to maintain the invariant.

It is easy to check that queue lengths do not exceed the number of elements printed.                                                                        □

The next problem is related to graphs (see chapter 9 for additional graph problems).

Let $V$ be a finite set whose elements are called *vertices*. Let $E$ be a subset of the set $V \times V$; the elements of $E$ are called *edges*. The sets $E$ and $V$ define a *directed graph*. A pair $\langle p, q \rangle \in E$ is called an edge *going from p to q*. One says that this edge *leaves p* and *enters q*. Usually vertices are drawn as points and edges as arrows. According to the above definition, there is at most one edge from $p$ to $q$; edges that are loops (from $p$ to $p$) are allowed.

A (directed) *path* is a sequence of vertices connected by edges (for example, path *pqrs* contains four vertices $p$, $q$, $r$, and $s$, connected by three edges $\langle p, q \rangle$, $\langle q, r \rangle$, and $\langle r, s \rangle$.

**6.2.6.** Suppose a directed graph satisfies two requirements: (1) it is connected; that is, there is a path from any given vertex to any other vertex; and (2) for any vertex the number of incoming edges is equal to the number of outgoing edges. Prove there exists an edge cycle that traverses each edge exactly once. Give an algorithm to find this cycle.

*Solution.* A "worm" is a nonempty queue of vertices such that each pair of adjacent vertices is connected by a graph edge (going in the direction from the first element to the last element). The first element in the queue is the "tail" of the worm; the last element in the queue is the worm's "head". The worm can be drawn as a chain of arrows; arrows lead from the tail to the head. When a vertex is added, the worm grows near the head; when a vertex is removed, the tail is cut off.

Initially, the worm consists of a single vertex. It evolves according to the following rule:

```
while the worm includes not all the edges do begin
  if there is an unused edge leaving the worm's head
      then begin
    add this edge to the worm
  end else begin
    {the head and tail of the worm are the same vertex}
    cut a piece of tail and add it to the head
    {"the worm eats its own tail"}
  end;
end;
```

Let us prove that this algorithm terminates when the worm spans all edges with its head and tail at the same vertex.

(1) Traversing the worm from tail to head, we enter each vertex as many times as we leave it. We also know that each vertex has as many incoming edges as it has outgoing edges. Therefore, we fail to find an outgoing edge only if the head of the worm is located at the same vertex as its tail.

(2) The worm never becomes shorter. Therefore, it will eventually reach some maximal length and never grow again. In the latter case, the worm will slide over itself forever. This is possible only if all the vertices visited do not have free outgoing edges. Since the graph is connected, this is possible only if all the edges are included in the worm.

Some remarks about the Pascal implementation. The vertices are numbered $1..n$. For each vertex $i$, we store the number $Out[i]$ of outgoing edges, as well as the numbers $Num[i][1], \ldots, Num[i][Out[i]]$ of vertices receiving the outgoing edges. While constructing the worm, we always choose the first unused edge. In this case, it is enough to keep (for each vertex) only the *number* of used outgoing edges to find the first unused edge. □

**6.2.7.** Prove that for any $n$ there exists a bit string $x$ of length $2^n$ with the following property: any binary string of length $n$ is a substring of the string $xxx \ldots$. Find an algorithm that constructs such a binary string in time $C^n$ (for some constant $C$ that does not depend on $n$).

[*Hint.* Consider a graph whose vertices are binary strings of length $n - 1$. An edge leaving $x$ and entering $y$ exists if and only if there is a string $z$ of length $n$ such that $x$ is a prefix of $z$ and $y$ is a suffix of $z$. (In other words, if $x$ minus its first bit is equal to $y$ minus its last bit.) This graph is connected; each vertex has two incoming and two outgoing edges. A cycle that traverses all edges provides a string satisfying the desired property.] □

**6.2.8.** Implement $k$ queues with total length not exceeding $n$, using memory of size $O(n+k)$ (that is, not exceeding $C(n+k)$ for some constant $C$). Each operation (except for initialization, which makes all the queues empty) should run in time $O(1)$ (that is, limited by a constant that does not depend on $n$).

*Solution.* We use the same method as for the pointer implementation of stacks. For each queue, remember the element that is first to be served; each element of the queue remembers the next element (the one that came immediately after). The last element believes that the next one is a special element number 0. We also have to remember the last element of each queue (otherwise we would trace the queue each time when a new element is added). As for stacks, all the free places are linked into a chain. Please note that for an empty queue the information about the last element makes no sense and is not used when adding elements.

```
Content: array [1..n] of T;
Next: array [1..n] of 0..n;
First: array [1..k] of 0..n;
Last: array [1..k] of 0..n;
Free: 0..n;

procedure Make_empty;
  var i: integer;
begin
  for i := 1 to n-1 do begin
    Next [i] := i + 1;
  end;
  Next [n] := 0;
  Free := 1;
  for i := 1 to k do begin
    First [i]:=0;
  end;
end;

function Is_space: Boolean;
begin
  Is_space := Free  <> 0;
end;

function Is_empty (queue_number: integer): Boolean;
begin
  Is_empty := First [queue_number] = 0;
end;

procedure Take (var t: T; queue_number: integer);
  var frst: integer;
begin
  {not Is_empty (queue_number)}
```

```
    frst := First [queue_number];
    t := Content [frst];
    First [queue_number] := Next [frst];
    Next [frst] := Free;
    Free := frst;
end;

procedure Add (t: T; queue_number: integer);
  var new, lst: 1..n;
begin
  {Is_space}
  new := Free; Free := Next [Free];
  {location new is removed from free space list}
  if Is_empty (queue__number) then begin
    First [queue_number] := new;
    Last [queue_number] := new;
    Next [new] := 0;
    Content [new] := t;
  end else begin
    lst := Last [queue_number];
    {Next [lst] = 0}
    Next [lst] := new;
    Next [new] := 0;
    Content [new] := t;
    Last [queue_number] := new;
  end;
end;

function First_element (queue_number: integer): T;
begin
  First_element := Contents [First [queue_number]];
end;                                                    □
```

**6.2.9.** The same problem for deques.

[*Hint*. A deque is a symmetric structure, so we should keep pointers to both the next and preceding elements. It is convenient to tie the ends of each deque with a special element forming a "ring". Another ring can be constructed from the free locations.]                                                    □

In the following problem, the deque is used to store the vertices of a convex polygon.

**6.2.10.** Assume that $n$ points in the plane are numbered from left to right (and when the $x$-coordinates coincide, according to the order of the $y$-coordinates). Write a program that finds the convex hull of these $n$ points in time $O(n)$ (that is, the number of operations should not exceed $Cn$ for some constant $C$). The convex hull is a polygon, so the answer should be a list of all its vertices.

*Solution.* Consider the points one by one, each time adding a new point to the existing convex hull. The ordering guarantees that the new point becomes one of the vertices of the convex hull. We call this vertex of the convex hull a "marked" vertex. At the next step the marked vertex is visible from the point to be added. We extend our polygon by a "needle", which goes from the marked vertex to the new point and back. We obtain a degenerate polygon and then eliminate "concavities" in that polygon.

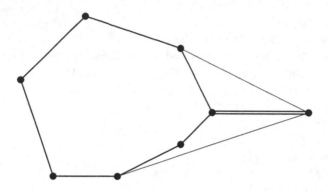

The program stores the vertices of a polygon in a deque listed counter-clockwise from the "head" to the "tail". The marked vertex is both the head and the tail of the deque. Adding a "needle" means that new vertex is added to both ends of the deque. The elimination of concavities is more difficult. Let us call the elements nearest the head the "subhead" and "subsubhead", respectively. The elimination of concavities near the head is done as follows:

```
while going from the head to the subsubhead we turn
    to the right near the subhead  do begin
  remove the subhead from the deque
end
```

The concavity near the tail is eliminated in a similar way.

*Remark.* Strictly speaking, operations involving the sub-head and "sub-sub-head" of a deque are not allowed by definition. However, they may be reduced to a few legal operations (for example, we can take three elements, process them, and put back what remains).

*Another remark*: Two degenerate cases are possible. The first occurs when we do not turn at all near the sub-head (in this case, the three vertices lie on the same line); the second occurs when we make a 180° turn (this happens when we have a "polygon" with two vertices). In the first case, the sub-head should be removed (to eliminate the redundant vertices from the convex hull); in the second case, the deque is left unchanged.                                                                 □

## 6.3 Sets

Let T be a type. There are several methods to store (finite) sets of values of type T. There is no "best" method; the choice depends on type T and on the operations needed.

**Subsets of $\{1, \ldots, n\}$**

**6.3.1.** Using $O(n)$ space (=space proportional to $n$), store a subset of $\{1, \ldots, n\}$.

| Operations | Time |
|---|---|
| Make empty | $Cn$ |
| Test membership | $C$ |
| Add | $C$ |
| Delete | $C$ |
| Minimal element | $Cn$ |
| Test if the set is empty | $Cn$ |

*Solution.* Store the set as `array [1..n] of Boolean`. □

**6.3.2.** The same problem with an additional requirement: test if the set is empty in constant (i.e., $O(1)$) time.

*Solution.* Store the number of elements in an additional variable. □

**6.3.3.** The same problem with the following restrictions:

| Operations | Time |
|---|---|
| Make empty | $Cn$ |
| Test membership | $C$ |
| Add | $C$ |
| Delete | $Cn$ |
| Minimal element | $C$ |
| Test if the set is empty | $C$ |

*Solution.* Maintain also the minimal element of the set. □

**6.3.4.** The same problem with the following restrictions:

| Operations | Time |
|---|---|
| Make empty | $Cn$ |
| Test membership | $C$ |
| Add | $Cn$ |
| Delete | $C$ |
| Minimal element | $C$ |
| Test if the set is empty | $C$ |

*Solution.* Store the minimal element of the set. Also, for each element we maintain pointers to the next and preceding elements (in order determined by value). □

**Sets of integers**

In the following problems, elements of the set are integers (unbounded); the number of elements does not exceed $n$.

**6.3.5.** The memory size is limited by $Cn$.

| Operations | Time |
|---|---|
| Make empty | $C$ |
| Cardinality | $C$ |
| Test membership | $Cn$ |
| Add element (known to be absent) | $C$ |
| Delete | $Cn$ |
| Minimal element | $Cn$ |
| Take some element | $C$ |

*Solution.* The set is represented by the variables

```
a:array [1..n] of integer, k: 0..n;
```

The set contains k (distinct) elements a[1],...,a[k]. In a sense, we keep the elements of the set in a stack. (We require all elements in the stack to be different.) We may also use a queue instead of a stack.                                                      □

**6.3.6.** The memory size is limited by $Cn$.

| Operations | Time |
|---|---|
| Make empty | $C$ |
| Test if the set is empty | $C$ |
| Test membership | $C \log n$ |
| Add | $Cn$ |
| Delete | $Cn$ |
| Minimal element | $C$ |

*Solution.* We use the same representation as in the preceding problem, with the additional restriction a[1] < ... < a[k]. To test membership, we use a binary search.                                                                                              □

In the following problem, different methods are combined.

**6.3.7.** Find all the vertices of a directed graph that can be reached from a given vertex along the graph edges. The program should run in time $Cm$, where $m$ is the total number of edges leaving the reachable vertices.

*Solution.* (See also a recursive solution in chapter 7.) Let num[i] be the number of outgoing edges for vertex i (assume that vertices are numbered 1..n). Let out[i][1],..., out[i][num[i]] be the endpoints of the edges starting from vertex i.

```
procedure Print_Reachable (i: integer);
    {print all the vertices reachable from i,
        including the vertex i itself}
var  X: subset of 1..n;
     P: subset of 1..n;
     q, v, w: 1..n;
     k: integer;
begin
  ...make X and P empty;
  writeln (i);
  ...add i to X, P;
  {(1) P is the set of printed vertices; P contains i;
   (2) only vertices reachable from i are printed;
   (3) X is a subset of P;
   (4) all printed vertices which have an outgoing edge
        to a non-printed vertex, belong to X}
  while X is not empty do begin
    ...take some element of X into v;
    for k := 1 to num [v] do begin
      w := out [v][k];
      if w does not belong to P then begin
        writeln (w);
        add w to P;
        add w to X;
      end;
    end;
  end;
end;
```

Let us check that the requirements (1)–(4) mentioned in the program text, are satisfied.

(1) We print a number and simultaneously add it to *P*.

(2) Since v is in X, v is reachable; therefore, w is reachable.

(3) Obvious.

(4) We delete v from X, but all the endpoints of edges emanating from v are then printed.

Let us prove the upper bound for the number of operations. If some element is removed from X, it never appears in X again. Indeed, it was present in P when removed, and only elements not in P can be added. Therefore the body of the while-loop is executed at most once for any reachable vertex; the number of iterations of the for-loop is equal to the number of outgoing edges.

For X we may use a stack or queue representation (see above); for P we use a Boolean array. □

The choice between stack and queue representation influences the order in which vertices are printed, as the following problem shows:

**6.3.8.** Solve the preceding problem if all the reachable vertices are to be printed in the following order: first the given vertex, then its neighbors, then (unprinted) neighbors of its neighbors, etc.

[*Hint.* Use a queue for the representation of the set X in the program above. By induction over $k$ we prove that at some point all the vertices having distance not exceeding $k$ (and no others) are printed, and all the vertices having distance exactly $k$ (and no others) are in the queue. For the detailed solution see section 9.2, p. 133.] □

More elaborate data structures for sets are considered in chapters 13 (hash tables) and 14 (trees).

## 6.4 Priority queues

**6.4.1.** Implement a data structure that has the same set of operations as an array of length n, namely,

- initialize;
- put x in the i-th cell;
- find the contents of the i-th cell;

as well as the operation

- find the index of the minimal element

(or one of the minimal elements). Any operation should run in time $C \log n$ (except for the initialization, which should run in time $Cn$).

*Solution.* We use the trick from the heapsort algorithm. Assume that the array elements are positioned at the leaves of a binary tree and each non-leaf vertex contains the minimum of its two sons. To maintain this information and to trace the path from the root to the minimal element, we need only $C \log n$ operations.                                   □

**6.4.2.** A *priority queue* does not employ First In First Out (FIFO) rule; only an element's priority is important. An element is added to the priority queue with some priority (which is assumed to be an integer). When an element is taken from the queue, it is the element with the greatest priority (or one of the elements with greatest priority). Implement a priority queue in such a way that adding and removing elements requires logarithmic (in the size of the queue) time.

*Solution.* Here we follow the idea of the heapsort algorithm in its final form. We place queue elements in an array x[1..k] and maintain the following invariant relation: x[i] is higher (has greater priority) than its sons x[2i] and x[2i+1], if they exist. (Therefore, each element is higher than all its descendants.) The priority information is maintained along with the elements in the array, so we have an array of pairs ⟨element, priority⟩. From the heapsort algorithm, we know how to delete an element and maintain this relation. Another thing we need to do is restore this relation after adding some element to the end of the array. This is done as follows:

```
    t:= the number of element added
    {invariant: any element is higher than any its
       descendant if the descendant is not t}
    while t is not root and t is higher than its father
     |    do begin
     |  exchange t and its father
    end;
```

Suppose the priority queue is formed by people standing at the vertices of a tree (drawn on the ground); each person has one predecessor and at most two successors. The idea of the algorithm is this: A highly-ranked individual added to the queue begins to move toward the head of the queue. If a predecessor has lower rank, this new individual takes the predecessor's place. This continues until a higher-ranked predecessor is encountered. □

*Remark.* The priority queue is an important data structure in simulation. Indeed, events are taken to be queue elements where the priority is determined by the time planned for the event.

# 7

# Recursion

Up to now, we have not use recursion in our examples. Instead we devote a special chapter to this important programming technique. Recursion can be very useful and convenient, and in some cases recursive solutions are much shorter and nicer than non-recursive ones.

We start with simple examples (section 7.1) to illustrate how recursive programs work. Then (section 7.2) we consider a class of problems where recursion is especially useful (tree processing). In section 7.3 we return to the problems considered in chapters 2 and 3 and show how they can be solved using recursion. Some other examples (topological sorting, finding connected components, etc.) are considered in section 7.4.

## 7.1 Examples

Let us start with general remarks. Assume a recursive procedure (that calls itself) is given, and we want to show that:

(a) the procedure terminates;
(b) the procedure works properly (assuming it terminates).

How can we do that? Let us start with (b). Here it is enough to check that a procedure containing a recursive call works properly assuming that the called program (with the same name) works properly. Indeed, in this case, all the programs in the chain of recursive calls (from the end of the chain to its beginning) work properly. In other words, a recursive procedure is proved correct by induction.

To prove (a) we usually find a parameter that decreases as the recursion depth increases and prove that it cannot decrease indefinitely.

**7.1.1.** Write a recursive program that computes the factorial of a positive integer $n$ (i.e., the product $n! = 1 \cdot 2 \cdots n$).

*Solution.* We apply the relations $1! = 1$ and $n! = (n - 1)! \cdot n$ for $n > 1$.

A. Shen, *Algorithms and Programming*, Springer Undergraduate Texts
in Mathematics and Technology, DOI 10.1007/978-1-4419-1748-5_7,
© Springer Science+Business Media, LLC 2010

```
procedure factorial (n: integer; var fact: integer);
| {fact := n!}
begin
  | if n=1 then begin
  | | fact:=1;
  | end else begin {n>1}
  | | factorial (n-1, fact);
  | | {fact = (n-1)!}
  | | fact:= fact*n;
  | end;
end;                                                                □
```

Using Pascal functions, we may write the above procedure as follows:

```
function factorial (n: integer): integer;
begin
  | if n=1 then begin
  | | factorial := 1;
  | end else begin {n>1}
  | | factorial := factorial(n-1)*n;
  | end;
end;
```

Please note that in this program the identifier factorial has two different meanings. It is a local variable as well as a function name. Fortunately, the difference is clear because the function name has parentheses after it. However, in the case of a function without parameters we have to be careful. (A common error occurs as the programmer sees a variable whereas the compiler sees a recursive call. This error is sometimes difficult to find.)

**7.1.2.** The factorial of 0 is defined as $0! = 1$ (note that $n! = n \cdot (n-1)!$ for $n = 1$). Modify the program accordingly.                                □

**7.1.3.** Write a recursive program that computes the nonnegative integral power of a real number.                                                        □

**7.1.4.** Repeat the previous problem with the requirement: the recursion depth (number of recursion levels) should not exceed $C \log n$, where $n$ is the exponent.

*Solution.*

```
function power (a,n: integer): integer;
begin
  | if n = 0 then begin
  | | power:= 1;
  | end else if n mod 2 = 0 then begin
  | | power:= power(a*a, n div 2);
  | end else begin
```

```
     |  | power:= power(a,n-1)*a;
     |  end;
     end;
```
□

**7.1.5.** What happens if we replace the line

```
power:= power(a*a, n div 2)
```

in the above program by the line

```
power:= power(a, n div 2)* power(a, n div 2)
```

*Solution.* The program is still correct, but becomes much slower. In this case, one call of the function power generates *two* calls of the same function (with identical parameters). Thus, the number of calls grows exponentially as a function of the recursion depth. The program still has logarithmic recursion depth, but the number of steps is now linear instead of logarithmic. □

This difficulty can be avoided by writing

```
t:= power(a, n div 2);
power:= t*t;
```

or by using Pascal's square function (sqr).

**7.1.6.** Using the procedure write(x) for x = 0...9, write a recursive procedure that prints the decimal representation of a positive integer *n*.

*Solution.* The recursive solution allows us to produce digits from right to left but print them from left to right:

```
procedure print (n:integer); {n>0}
begin
   | if n<10 then begin
   |    | write (n);
   | end else begin
   |    | print (n div 10);
   |    | write (n mod 10);
   | end;
end;
```
□

**7.1.7.** The "Towers of Hanoi" puzzle consists of three vertical sticks and *N* rings of different sizes. The rings are put on one of the sticks in such a way that larger rings are beneath smaller ones. We are to move this tower onto another stick one ring at a time. While moving the rings from one stick to another, we are not permitted to put a larger ring onto a smaller one. Write a program that shows the list of movements required to solve the problem.

*Solution.* The following recursive procedure moves i upper rings from the m-th stick to the n-th stick (we assume that the remaining rings on all sticks are larger and remain untouched); m and n are different numbers among {1, 2, 3}:

```
procedure move(i,m,n: integer);
 | var s: integer;
begin
 | if i = 1 then begin
 |  | writeln ('move ', m, '->', n);
 | end else begin
 |  | s:=6-m-n; {s is the third stick; 1+2+3 = 6}
 |  | move (i-1, m, s);
 |  | writeln ('move ', m, '->', n);
 |  | move (i-1, s, n);
 | end;
end;
```

(The first recursive call moves a tower of i-1 rings onto the third stick. After that the i-th ring becomes free and is moved to the remaining stick. The second recursive call moves the tower onto the i-th ring.)                                                □

**7.1.8.** Write a recursive program that computes the sum of all elements in an array a: array [1..n] of integer.

[*Hint.* A recursively defined function may have as a parameter the number of elements that should be added.]                                                □

## 7.2 Trees: recursive processing

*Reminder*: A binary tree is represented by a picture like this:

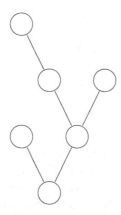

The vertex at the bottom of the tree is called the *root*. Two lines may go up from any vertex: one going up-left and one going up-right. These two vertices are called the *left* and *right sons* of the given vertex. Any given vertex may have either two sons, one son (which may be either the left son or the right son), or no sons at all. In the latter case, the vertex is called a *leaf*.

Let $x$ be a vertex of tree. Consider this vertex together with its sons, grandsons, etc. This is a *subtree rooted at $x$*, the subtree of all *descendants* of the vertex $x$.

Please note that in most textbooks trees have root at the top and grow downwards; terms "son", "father", "brother" are usually replaced by "child", "parent", "sibling", etc.

In the following set of problems tree vertices are numbered by positive integers, and all numbers are different. The number assigned to the tree root is kept in the variable root. There exist two arrays

        l,r: array [1..N] of integer

The left and right sons of the vertex number i have numbers l[i] and r[i], respectively. If vertex $x$ has no left (or right) son, the value of l[i] (resp., r[i]) is equal to 0. (Following the tradition, we use the symbolic constant nil instead of the literal 0.) Numbers of all vertices do not exceed N.

Let us stress that the vertex number has no connection with its position in a tree and that some integers in $1 \ldots N$ are not assigned to vertices at all. (Therefore, some data in the arrays l and r are irrelevant.)

**7.2.1.** Assume that N = 7, root = 3, and the arrays l and r are as follows:

| i | 1 | 2 | 3 | 4 | 5 | 6 | 7 |
|------|---|---|---|---|---|---|---|
| l[i] | 0 | 0 | 1 | 0 | 6 | 0 | 7 |
| r[i] | 0 | 0 | 5 | 3 | 2 | 0 | 7 |

Draw the corresponding tree.

Answer:

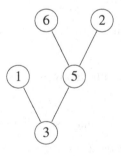

□

**7.2.2.** Write a program that counts all the vertices in a given tree.

*Solution.* Consider a function n(x), which is defined as the number of vertices in the subtree rooted at vertex number x. We agree that n(nil) = 0 (and the corresponding subtree is empty) and ignore the values n(s) for s not assigned to any vertex. The answer is n(root). Here is a recursive program that computes n(x):

```
function n(x:integer):integer;
begin
  if x = nil then begin
    n:= 0;
  end else begin
    n:= n(l[x]) + n(r[x]) + 1;
  end;
end;
```

(Vertices in the x-subtree are vertices in the subtrees rooted at its sons plus the vertex x itself.) The procedure terminates because the recursive calls refer to trees of smaller heights.                                                            □

**7.2.3.** Write a program that counts the leaves in a tree.

*Solution.*

```
function n (x:integer):integer;
begin
  if x = nil then begin
    n:= 0;
  end else if (l[x]=nil) and (r[x]=nil) then begin {leaf}
    n:= 1;
  end else begin
    n:= n(l[x]) + n(r[x]);
  end;
end;
```
                                                                            □

**7.2.4.** Write a program that finds the height of a tree. (The root of a tree has depth 0, its sons have depth 1, its grandsons have depth 2, etc. The height of a tree is the maximal depth of its vertices.)

[*Hint*. Let h(x) be the height of the subtree rooted at x. The function h may be defined recursively.]                                                        □

**7.2.5.** Write a program which for a given n counts all the vertices of depth n in a given tree.                                                              □

Instead of counting vertices, we may ask to list them (in some order).

**7.2.6.** Write a program that prints all vertices (one time each).

*Solution.* The procedure print_subtree(x) prints all the vertices of the subtree rooted at x (each vertex is printed once). The main program consists of the call print_subtree(root).

```
procedure print_subtree (x:integer);
begin
  if x = nil then begin
```

```
    |  | {nothing to do}
    | end else begin
    |  | writeln (x);
    |  | print_subtree (l[x]);
    |  | print_subtree (r[x]);
    |  | end;
    | end;
```

This program uses the following ordering of tree vertices: first the root, then the left subtree, and then the right subtree. This order is determined by the order of the three lines in the else-part. Any of six possible permutations of these lines gives a specific order of tree traversal. □

## 7.3 The generation of combinatorial objects; search

Recursion is a convenient tool to write programs that generate elements of some finite set. As an example, we now return to the problems of chapter 2.

**7.3.1.** Write a program that prints all sequences of length n composed of the numbers 1..k. (Each sequence should be printed once, so the program prints $k^n$ sequences.)

*Solution.* The program employs an array a[1]..a[n] and an integer variable t. The recursive procedure generate prints all sequences with prefix a[1]..a[t]; after it terminates, the value of t and a[1]..a[t] are the same as before the call.

```
    procedure generate;
    | var i,j : integer;
    begin
    | if t = n then begin
    |  | for i:=1 to n do begin
    |  |  | write(a[i]);
    |  | end;
    |  | writeln;
    | end else begin {t < n}
    |  | for j:=1 to k do begin
    |  |  | t:=t+1;
    |  |  | a[t]:=j;
    |  |  | generate;
    |  |  | t:=t-1;
    |  | end;
    | end;
    end;
```

The main program body now consists of two lines:

```
    t:=0;
    generate;
```

*Remark.* For efficiency reasons we may move the commands t:=t+1 and t:=t-1 out of the for-loop.                                                                           □

**7.3.2.** Write a program that prints all permutations of the numbers 1..n (each should be printed once).

*Solution.* The program utilizes an array a[1]..a[n] that contains a permutation of numbers 1..n. The recursive procedure generate prints all permutations that have the same first t elements as the permutation a. After the call, the values of t and a are the same as before the call. The main program is:

```
for i:=1 to n do begin
|  a[i]:=i;
end;
t:=0;
generate;
```

The procedure definition is as follows:

```
procedure generate;
|  var i,j : integer;
begin
  if t = n then begin
    for i:=1 to n do begin
    |  write(a[i]);
    end;
    writeln;
  end else begin {t < n}
    for j:=t+1 to n do begin
      ..exchange a[t+1] and a[j]
      t:=t+1;
      generate;
      t:=t-1;
      ..exchange a[t+1] and a[j]
    end;
  end;
end;
```
                                                                           □

**7.3.3.** Print all sequences of length $n$ that contain $k$ ones and $n - k$ zeros. (Each of them should be printed once.)                                                   □

**7.3.4.** Print all increasing sequences of length k constructed from the natural numbers 1..n. (We assume that k ≤ n; otherwise the sequences do not exist.)

*Solution.* The program utilizes an array a[1]..a[k] and integer variable t. Assuming that a[1]..a[t] is an increasing sequence whose terms are numbers in 1..n, the recursive procedure generate prints all its increasing extensions of length k. (After the call, the values of t and a[1]..a[t] are the same as before the call.)

```
procedure generate;
| var i: integer;
begin
| if t = k then begin
| | ...print a[1]..a[k]
| end else begin
| | t:=t+1;
| | for i:=a[t-1]+1 to t-k+n do begin
| | | a[t]:=i;
| | | generate;
| | end;
| | t:=t-1;
| end;
end;
```

*Remark.* The for-loop may use n instead of $t - k + n$. The above version is more efficient; we use that the (k-1)-th term cannot exceed n-1, the (k-2)-th term cannot exceed n-2, etc.

The main program:

```
t:=1;
for j:=1 to 1-k+n do begin
| a[1]:=j;
| generate;
end;
```

(Another possibility is to add to a an auxiliary element a[0]:=0, then let t:=0 and call the procedure generate once.) □

**7.3.5.** Generate all representations of a given positive integer n as the sum of a non-increasing sequence of positive integers.

*Solution.* The program uses an array a[1..n] (the maximal number of summands is n) and an integer variable t. The procedure generate assumes that a[1]...a[t] is a non-increasing sequence of positive integers whose sum does not exceed n, and prints all the representations that extend this sequence. For efficiency reasons, the sum $a[1] + \cdots + a[t]$ is kept in an auxiliary variable s.

```
procedure generate;
| var i: integer;
begin
| if s = n then begin
| | ...print a[1]..a[t]
| end else begin
| | for i:=1 to min(a[t], n-s) do begin
| | | t:=t+1;
| | | a[t]:=i;
| | | s:=s+i;
```

```
        generate;
        s:=s-i;
        t:=t-1;
      end;
    end;
end;
```

The main program looks like

```
t:=1;
for j:=1 to n do begin
  a[1]:=j
  s:=j;
  generate;
end;
```

*Remark.* A small improvement is possible, since we may move the statements that increase and decrease t out of the loop. Also, instead of setting the value of s each time (s:=s+i) and restoring it (s:=s-i) we may increase it by 1 at each time through the loop and restore the original value at the end of loop. Finally, we may add an auxiliary element a[0] = n and simplify the main program:

```
t:=0; s:=0; a[0]:=n; generate;
```
□

**7.3.6.** Write a recursive program that traverses a tree (using the same statements and conditions as in chapter 3).

*Solution.* The procedure process_above processes all the leaves above the robot's position and returns the robot to the start position. Here is the recursive definition:

```
procedure process_above;
begin
  if is_up then begin
    up_left;
    process_above;
    while is_right do begin
      right;
      process_above;
    end;
    down;
  end else begin
    process;
  end;
end;
```
□

## 7.4 Other applications of recursion

**Topological sorting**. Imagine $n$ government officials, each of whom issues permissions of some type. We wish to obtain all the permissions (one from each official) according to the rules. The rules state (for each official) a list of permissions that should be collected before you can obtain this permission. There is no hope of solving the problem if the dependency graph has a cycle (we cannot get permission from $A$ without having $B$'s permission in advance, $B$ without $C, \ldots, Y$ without $Z$, and $Z$ without $A$). Assuming that such a cycle does not exist, we wish to find a plan that secures one of the permitted orders.

Let us represent officials by points and dependencies by arrows. (If permission $B$ should be obtained before $A$, draw an arrow going from $A$ to $B$.) We then have the following problem. There are $n$ points numbered from 1 to $n$. From each point there are several (maybe zero) arrows that go to other points. (This picture is called a *directed graph*.) The graph has no cycles. We want to put the graph vertices in such an order that the end of any arrow precedes its beginning. This is the problem of *topological sorting*.

**7.4.1.** Prove that it is always possible to topologically sort a directed graph without cycles.

*Solution.* The absence of cycles implies that there exists a vertex with no outgoing edges (otherwise, we can follow edges until we come to the already visited vertex). This vertex with no outgoing edges gets number 1. After we discard this vertex and all incident edges, we reduce our problem to the same problem with a smaller number of vertices. $\square$

**7.4.2.** Assume that a directed graph without cycles is stored in the following manner: Its vertices are numbered 1..n. For any i in 1..n, the value of num[i] is the number of edges leaving vertex i, and adr[i][1],..., adr[i][num[i]] are the numbers of vertices those edges enter. Write a (recursive) algorithm that performs a topological sort in time $C \cdot (n + m)$, where m is the number of edges (arrows) in the graph.

*Remark.* The solution to the preceding problem does not provide such an algorithm directly; we need a more ingenious construction.

*Solution.* Our program prints the vertices in question (their numbers). It uses an array

```
printed: array[1..n] of Boolean;
```

such that printed[i] is true if and only if vertex i is already printed (this information is updated each time a vertex is printed). We say that a sequence of printed vertices is *correct* if (**a**) no vertex is printed twice, and (**b**) for any printed vertex i all the edges leaving i enter the vertices that are printed before i.

```
procedure add (i: 1..n);
  {before: the sequence of printed vertices is correct}
  {after: the sequence of printed vertices is correct
```

```
|                    and includes i}
begin
|  if printed [i] then begin {i is printed already}
|  |  {nothing to do}
|  end else begin {printed sequence is correct}
|  |  for j:=1 to num[i] do begin
|  |  |  add(adr[i][j]);
|  |  end;
|  |  {printed sequence is correct; all the edges going out
|  |    of i are entering the vertices already printed; thus,
|  |    we may print i correctly if it is not printed yet}
|  |  if not printed[i] then begin
|  |  |  writeln(i); printed [i]:= TRUE;
|  |  end;
|  end;
end;
```

The main program is:

```
for i:=1 to n do begin
|  printed[i]:= FALSE;
end;
for i:=1 to n do begin
|  add(i)
end;
```

The time bound will be proven shortly.

**7.4.3.** The program above remains correct if we remove the test, replacing

```
if not printed[i] then begin
|  writeln(i); printed [i]:= TRUE;
end;
```

by

```
writeln(i); printed [i]:= TRUE;
```

Why? How should we change the specification of the procedure?

*Solution.* The specification of the procedure is now as follows:

```
{before: the sequence of printed vertices is correct}
{after: the sequence of printed vertices is correct
       and includes i; all newly printed vertices
       can be reached from i}
```
□

**7.4.4.** The correctness of the program depends on the assumption about the absence of cycles. However, we did not mention this assumption in the solution of problem 7.4.2. Why?

*Solution.* We omitted the proof that the program terminates. Let us give it now. For any vertex we define its *level* as the maximal length of a path going out of it along the edges. The level is finite because there are no cycles. Vertices of level 0 have no outgoing edges. For any edge the level of its endpoint is smaller than the level of its starting point by at least 1. When add(i) is executed, all recursive calls refer to vertices whose levels are smaller. □

Now we return to the time bound. How many calls add(i) are possible for some fixed i? The first call prints i; all others check that i is printed and exit immediately. It is also clear that all the calls add(i) are induced by the *first* calls of add(j) for all j such that the edge from j to i is present in the graph. Therefore, the number of calls add(i) is equal to the number of incoming edges for vertex i. All the calls except the first one require $O(1)$ time. The first requires time proportional to the number of outgoing edges (if we ignore the time needed to perform add(j) for endpoints of outgoing edges). Therefore the total time is proportional to the total number of edges (plus the number of vertices). □

**Connected component of a graph**. An *undirected graph* is a set of points (vertices) some of which are connected by lines (edges). An undirected graph can be considered as a special case of a directed graph where for each edge there is another edge going in the reverse direction.

The *connected component* of vertex i is the set of all vertices that are reachable from i via graph edges. Since the graph is undirected, the relation "j belongs to the connected component of i" is an equivalence relation.

**7.4.5.** Suppose an undirected graph is given (for each vertex its neighbors are listed; see the problem about topological sorting for details). Write an algorithm that for a given i prints all the vertices of the connected component of i (each vertex is printed once; no other vertices should be printed). The number of operations should be proportional to the total number of vertices and edges in the connected component.

*Solution.* The program will "blacken" vertices of the graph as they are printed. (Initially the vertices are assumed to be white.) By "white part" of the graph we mean that part of the graph that remains after we remove all black vertices and all edges adjacent to black vertices. The procedure add(i) blackens the connected component of i *in the white part of the graph* (and does nothing if i is already black).

```
procedure  add (i:1..n);
begin
  if i is black then begin
  | {nothing to do}
  end else begin
  | ..print i and mark i as black
  | for all j that are neighbors of i do begin
  | | add(j);
  | end;
  end;
end;
```

Let us prove that this procedure works properly (assuming that all recursive calls work properly). Indeed, it cannot blacken anything except the connected component of i. Let us check that all vertices in the connected component are blackened (and printed). Let k be a vertex that is reachable from x via path $i \rightarrow j \rightarrow \cdots \rightarrow k$, which includes only white vertices (and goes along graph edges). We may assume without loss of generality that this path does not visit vertex i again. Among all the paths with this property, we consider the path with the smallest j (in the order they are considered in the for-loop). Then after the calls add(m) for preceding neighbors m, no one of the vertices in the path $j \rightarrow \cdots \rightarrow k$ becomes black; otherwise, such a vertex (and therefore k) is white-reachable from m and j is not minimal. Therefore, k belongs to the connected component of the white part at the time when add(j) is called.

To prove that the algorithm terminates, it is enough to mention that the number of white vertices decreases at each recursion level.

Let us estimate the number of operations. Each vertex is blackened at most once, during the first call add(i) (for a given i). All subsequent calls occur when one of the neighbors is blackened. Therefore, the number of those calls is limited by the number of neighbors. And the only thing that happens during those calls is the check that i is already black. On the other hand, during the first call to add(i) all neighbors are considered and corresponding recursive calls are made. Therefore the total number of operations related to vertex i (not including the operations performed during the recursive calls add(j) for its neighbors j) is proportional to the number of neighbors of i. The upper bound stated in the problem follows.                          □

**7.4.6.** Solve the same problem for a directed graph (that is, print all the vertices accessible from a given vertex). Note: the graph may contain cycles.

*Solution.* Essentially the same program can be used. The line "for all neighbors of a vertex" should be replaced by "for all endpoints of outgoing edges".          □

The following version of the connected component problem is of theoretical importance (though its practical value is minimal). The statement is called the *Savitch theorem*.

**7.4.7.** A directed graph has $2^n$ vertices indexed by $n$-bit strings. It is presented as a function Edge_exists which for any two vertices $x$ and $y$ says whether there is an edge going from $x$ to $y$ or not. Write an algorithm that for given vertices $s$ and $t$ says whether there is a path from $s$ to $t$ and uses polynomial (in $n$) amount of memory. Computation time is not restricted.

[*Hint.* Write a recursive procedure that gets strings $u$, $v$ and an integer $k$ and says whether there is a path from $u$ to $v$ that has length at most $2^k$. This procedure calls itself with parameter $k - 1$ instead of $k$.]          □

**Hoare Quicksort.** A well-known sorting algorithm called "quicksort" is a recursive algorithm considered to be one of the fastest algorithms available. Assume that an array a[1]..a[n] is given. The recursive procedure sort(l,r:integer) sorts an interval of the array a with indices in (l, r]; that is, a[l+1]..a[r] (leaving the remaining part of the array unchanged).

```
procedure sort (l,r: integer);
begin
  if l = r then begin
    | {nothing to do - the interval is empty}
  end else begin
    ..find a random number s in the interval (l,r]
    b := a[s]
    ..rearrange the elements of the interval (l,r]
          into three parts:
      the elements smaller than b  - the interval (l,ll]
      the elements equal to b      - the interval (ll,rr]
      then elements greater than b - the interval (rr,r]
    sort (l,ll);
    sort (rr,r);
  end;
end;
```

How do we rearrange the elements of the interval according to the three categories listed in the above algorithm? As problem 1.2.32 (p. 27) shows, it can be done in time proportional to the length of the interval. Termination is guaranteed because the length of the interval decreases by at least 1 for each recursion level.

**7.4.8.** (For readers familiar with probability theory) Prove that the expected number of operations of the Hoare quicksort algorithm does not exceed $Cn \log n$ where the constant $C$ does not depend on the array to be sorted.

[*Hint.* Let $T(n)$ be the maximal value of the expected number of operations (the maximal is taken over all possible inputs of length $n$). The following inequality holds:

$$T(n) \leqslant Cn + \frac{1}{n} \sum_{k+l=n-1} \left( T(k) + T(l) \right)$$

Indeed, the first summand corresponds to the stage where all elements are rearranged (divided into "less than", "equal to", or "greater than" parts). The second summand is an average value taken over all possible choices of a random number s. (To be precise, we should note that some of the elements may be equal to the threshold, so instead of $T(k)$ and $T(l)$ we should use the maximum of $T(x)$ over all $x$ not exceeding $k$ (or $l$), but this makes no difference.)

Now, we prove by induction over $n$ that $T(n) \leqslant C'n \ln n$. To compute the average value of $x \ln x$ for all integer $x$ such that $1 \leqslant x \leqslant n-1$, we integrate $\int_1^n x \ln x \, dx$ by parts as $\int \ln x \, d(x^2)$. When $C'$ is large enough, the term $Cn$ on the right-hand side is absorbed by the integral $\int x^2 \, d \ln x$ and the inductive step is finished.]   □

**7.4.9.** An array of $n$ different integers and a number $k$ is given. Find the $k$-th element of the array (in increasing order) using at most $Cn$ operations (where $C$ is some constant that does not depend on $k$ and $n$).

*Remark.* Sorting algorithms can be used, but the number of operations ($Cn \log n$) is too big. The naïve algorithm (find the minimal element, then the next one, ..., then the $k$-th one) requires about $kn$ operations (which is not allowed, because the constant $C$ must not depend on $k$).

[*Hint.* An elegant method (though hardly practical since the constants are rather big) goes as follows:

A. Separate the array into $n/5$ groups each containing 5 elements. Sort each group.

B. Consider the median (third) elements of each group. This gives an array of $n/5$ elements. Calling our algorithm recursively, find the median element of this array. Call it $M$.

C. Compare all other elements of the initial array with $M$. They are divided into two groups (elements less than $M$ and elements greater than $M$). Count the elements in both groups. Then we know which category the required ($k$-th) element belongs to and what its number is inside that part.

D. Apply the algorithm recursively to that part to find the required element.

Let $T(n)$ be the maximal possible number of operations when this algorithm is applied to arrays of length not exceeding $n$ (the number $k$ may be arbitrary). We have the following bound:

$$T(n) \leqslant Cn + T(n/5) + T(0.7n).$$

The last term may be explained as follows. Each of the three categories contains at least $0.3n$ elements. Indeed, about half of all the median elements (in 5-element sets) are smaller than $M$. And if a median element of a 5-element set is smaller than $M$, then at least two more elements are smaller than $M$. Therefore, 3/5 of half of all elements are smaller than $M$.

Now the bound $T(n) \leqslant Cn$ can be proved by induction. The crucial point is that $1/5 + 0.7 < 1.$ ]                                                                    □

# 8

# Recursive and non-recursive programs

For a universal programming language (like Pascal) recursion is, in a sense, redundant: for any recursive program it is possible to write an equivalent program without recursion. Of course, this does not mean that recursion should be avoided, because it allows us to provide elegant solutions to otherwise complicated problems.

However, we want to show some methods that allow us to eliminate recursion in some cases and transform a recursive program into an equivalent non-recursive program.

What for? A pragmatic answer is that sometimes recursion is implemented in a non-efficient way and recursive programs may be significantly slower than equivalent non-recursive programs. Another problem is that some programming languages do not allow recursion at all. But the main reason is that elimination of recursion is sometimes very instructive. In section 8.1 we describe a technique that often allows us not only to eliminate recursion, but also get a faster program. Then in section 8.2 we consider a more general approach. Finally, is section 8.3 we show some examples not covered by these techniques.

## 8.1 Table of values (dynamic programming)

**8.1.1.** The following recursive procedure computes binomial coefficients. Write an equivalent program without recursion.

```
function C(n,k: integer):integer;
| {n >= 0; 0 <=  k <=n}
begin
| if (k = 0) or (k = n) then begin
| | C:=1;
| end else begin {0<k<n}
| | C:= C(n-1,k-1)+C(n-1,k)
| end;
end;
```

A. Shen, *Algorithms and Programming*, Springer Undergraduate Texts
in Mathematics and Technology, DOI 10.1007/978-1-4419-1748-5_8,
© Springer Science+Business Media, LLC 2010

*Remark.* $C(n, k) = \binom{n}{k}$ is the number of $k$-element subsets of an $n$-element set. The identity

$$\binom{n}{k} = \binom{n-1}{k-1} + \binom{n-1}{k}$$

is proved as follows: Fix some element $x$ of the $n$-element set. Then all $k$-element subsets are divided into two categories: those that contain $x$ and those that do not. The elements of the first type are in one-to-one correspondence with the $(k-1)$-element subsets of a $(n-1)$-element set (just discard $x$); the elements of the second type are $k$-element subsets of a $(n-1)$-element set.

The table of $\binom{n}{k}$-values

$$
\begin{array}{ccccccccc}
 &  &  &  & 1 &  &  &  & \\
 &  &  & 1 &  & 1 &  &  & \\
 &  & 1 &  & 2 &  & 1 &  & \\
 & 1 &  & 3 &  & 3 &  & 1 & \\
1 &  & 4 &  & 6 &  & 4 &  & 1 \\
\end{array}
$$

. . . . . . . . . . . .

is called the *Pascal triangle* (the same Blaise Pascal who gave his name to a programming language). In this triangle, any element (except the 1s on the left and the right) is the sum of the two elements above it.

*Solution.* One may use the formula

$$\binom{n}{k} = \frac{n!}{k!\,(n-k)!}$$

We do not use it because we want to show more general methods to eliminate recursion. Our program fills the table of values $C(n,k) = \binom{n}{k}$ for $n = 0, 1, 2, \ldots$ until it reaches the element in question.                                                   □

**8.1.2.** Compare the computation time for the recursive and non-recursive versions of the binomial coefficient algorithm, and similarly for the amount of memory used.

*Solution.* The table used in the non-recursive version occupies space of order $n^2$. We can reduce it to $n$ if we recall that only one line of the Pascal triangle is needed to compute the next line. The time required is still $n^2$.

The recursive program requires much more time. Indeed, the call $C(n,k)$ causes two calls of type $C(n-1,..)$, those two calls cause four calls of type $C(n-2,..)$, etc. Hence, the time is exponential (of order $2^n$). The recursive procedure uses $O(n)$ memory (we have to multiply the recursion depth, that is $n$, by the amount of memory required by one copy of the procedure, that is $O(1)$).                                   □

The reason why the non-recursive version is so much faster is the following. In the recursive version, the same computations are repeated many times. For example, the call $C(5,3)$ causes *two* calls of $C(3,2)$:

```
        C(5,3)
       ↙    ↘
   C(4,2)   C(4,3)
   ↙   ↘   ↙   ↘
C(3,1)  C(3,2)  C(3,3)
```

When we fill the table, we compute the value for each cell only once, hence the economy of the non-recursive method. This method is called *dynamic programming*, and is useful when the amount of information to be stored in the table is not too large.

**8.1.3.** Compare the recursive and the (simplest) non-recursive algorithm to compute the Fibonacci numbers defined as the sequence

$$\Phi_1 = \Phi_2 = 1; \quad \Phi_n = \Phi_{n-1} + \Phi_{n-2} \ (n > 2).\qquad\square$$

**8.1.4.** A convex polygon with $n$ vertices is given (by a list of coordinates of its vertices). It is cut into triangles by non-intersecting diagonals. To do this, we need exactly $n - 3$ diagonals (proof by induction over $n$). The cost of the triangulation is defined as the total length of all the diagonals used. Find the minimal cost of the triangulation. The number of operations should be limited by a polynomial of $n$. (This requirement excludes exhaustive search, since the number of possibilities is not bounded by any polynomial.)

*Solution.* Assume that the vertices are numbered from 1 to $n$ and the numbers increase in the clockwise direction. Let $k$ and $l$ be two numbered vertices and assume $l > k$. By $A(k, l)$ we denote a polygon cut from the original polygon by segment $k$–$l$. (The segment $k$–$l$ cuts the polygon into two polygons, one of which contains the 1–$n$ side; $A(k, l)$ is the other one.) The initial polygon is denoted by $A(1, n)$. When $l = k + 1$, we have a degenerate polygon with only two vertices.

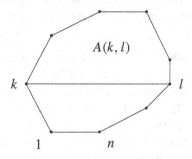

By $a(k, l)$ we denote the minimal cost of triangulation of $A(k, l)$. Let us write a recurrence formula for $a(k, l)$. When $l = k + 1$, we have a degenerate polygon with two vertices and let $a(k, l) = 0$. When $l = k + 2$, we have a triangle, and $a(k, l) = 0$. Assume that $l > k + 2$.

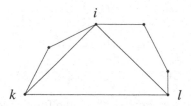

The chord $k$–$l$ is a side of the polygon $A(k, l)$; therefore, it is a side of one of the triangles of the triangulation. The opposite vertex of this triangle has some number $i$. It may be any of the vertices $k + 1, \ldots, l - 1$, and the minimal triangulation cost can be computed as

$$\min\{(\text{the length of } k\text{–}i) + (\text{the length of } i\text{–}l) + a(k, i) + a(i, l)\}$$

where the minimal value is taken over all $i = k + 1, \ldots, l - 1$. We should also take into account that for $q = p + 1$, the segment $p$–$q$ is one of the sides and its length should be counted as 0 for our purposes.

This formula allows us to fill the table of values $a(k, l)$ in order of increasing number of vertices (which is $l - k + 1$). The corresponding program uses memory of order $n^2$ and time of order $n^3$ (one application of the recurrent formula requires a search for a minimal value among not more than $n$ numbers).                            □

**8.1.5.** An $m \times n$ matrix is a rectangular table with $m$ rows and $n$ columns filled with numbers. An $m \times n$ matrix may be multiplied by an $n \times k$ matrix (the width of the left factor must be equal to the height of the matrix on the right) giving $m \times k$ matrix as the result. The cost of such a multiplication is defined as $mnk$ (this is the number of multiplications required by the standard multiplication algorithm, but this is not important). Matrix multiplication is associative, therefore the product of $s$ matrices may be computed in any order. For each ordering, consider the total cost of all matrix multiplications. Find the minimal cost when the sizes of the matrices are given. The running time of the algorithm should be bounded by a polynomial over the number of factors ($s$).

Example. Matrices of size $2 \times 3$, $3 \times 4$, $4 \times 5$ can be multiplied in two different ways. The cost is either $2 \cdot 3 \cdot 4 + 2 \cdot 4 \cdot 5 = 64$ or $3 \cdot 4 \cdot 5 + 2 \cdot 3 \cdot 5 = 90$.

*Solution.* Suppose the first matrix is associated with an interval $[0, 1]$, the second one is associated with $[1, 2], \ldots$, and the $s$-th matrix is associated with $[s - 1, s]$. Adjacent matrices (for segments $[i - 1, i]$ and $[i, i + 1]$) have a common dimension so we can multiply them. Let us denote this common dimension by $d[i]$. Therefore, the initial data of our problem is an array $d[0], \ldots, d[s]$.

Let $a(i, j)$ be the minimal cost of computation of the product of all the matrices in the interval $[i, j]$ (here $0 \leqslant i < j \leqslant s$). The cost in question is $a(0, s)$. The values of $a(i, i + 1)$ are equal to 0 (we have only one matrix and nothing to multiply). The recurrence formula is as follows:

$$a(i, j) = \min\{a(i, k) + a(k, j) + d[i]d[k]d[j]\}$$

where the minimal value is computed over all possible places of the last multiplication; that is, over all $k = i + 1, \ldots, j - 1$. Indeed, the product of all matrices in the interval $[i, k]$ is a matrix of size $d[i] \times d[k]$, the product of all the matrices in the interval $[k, j]$ has size $d[k] \times d[j]$, and the cost of multiplication is $d[i]d[k]d[j]$. □

*Remark.* The last two problems are rather similar. This is clear if we associate matrix factors with the sides 1–2, 2–3,...,$(s − 1)$–$s$ of a polygon, and associate any chord $i$–$j$ with the product of all matrices covered by this chord.

**8.1.6.** A one-way railway has $n$ stops. We know the price of tickets from the $i$-th stop to the $j$-th stop (for $i < j$, since there is no traffic in the other direction). Find the minimal travel cost from stop 1 to stop $n$ (taking into account possible savings due to intermediate stops). □

We have seen that sometimes we get a more effective algorithm by replacing the recursion with a table that is filled cell by cell. A similar effect is achieved if we retain the recursive algorithm, but store the values of the function already computed and do not compute them again when the second request occurs. This trick is called *memoization*.

**8.1.7.** A finite set and a binary operation $\langle u, v \rangle \mapsto u \circ v$ defined on this set are given (the operation may be noncommutative and nonassociative). We have $n$ elements $a_1, \ldots, a_n$ from the set and one more element $x$. Check if it is possible to insert parentheses in the expression $a_1 \circ \cdots \circ a_n$ in such a way that the result is equal to $x$. The number of operations should not exceed $Cn^3$ for some constant $C$ (which depends on the cardinality of the set given).

*Solution.* Fill a table that contains (for any subexpression $a_i \circ \cdots \circ a_j$) the set of all possible values (for different placements of parentheses). □

The same trick is used in the polynomial algorithm that tests whether a given word belongs to a context-free language (see section 15.1, p. 224).

The next problem (knapsack problem) was mentioned in section 3.4, p. 59.

**8.1.8.** An array $x_1, \ldots, x_n$ of $n$ positive integers and an integer $N$ are given. Check if $N$ is equal to the sum of some subset of $\{x_1, \ldots, x_n\}$. The number of operations should be of order $Nn$.

[*Hint.* After $i$ iterations, keep the set of all numbers in $0, \ldots, N$ that can be represented as a sum of some subset of $\{x_1 \ldots x_i\}$.] □

## 8.2 Stack of postponed tasks

We illustrate another way to eliminate recursion using the Towers of Hanoi (p. 105) problem as an example.

**8.2.1.** Write a non-recursive program that prints the sequence of moves for Towers of Hanoi problem.

*Solution.* Recall the following recursive program that moves i upper rings from stick m to stick n:

```
procedure move(i,m,n: integer);
| var s: integer;
begin
  | if i = 1 then begin
  |   | writeln ('move ', m, '->', n);
  | end else begin
  |   | s:=6-m-n; {s is the unused stick; 1+2+3=6}
  |   | move (i-1, m, s);
  |   | writeln ('move ', m, '->', n);
  |   | move (i-1, s, n);
  | end;
end;
```

This program reduces the task "move i rings from m to n" to three tasks of the same type. Two of them deal with i-1 rings; one of them deals with 1 ring.

Try to execute this program manually. You'll see that it is rather difficult to remember which tasks are still to be done on different recursion levels.

The non-recursive program uses a *stack of postponed tasks*, whose elements are triples $\langle i, m, n \rangle$. Each triple is interpreted as the request "move i (upper) rings from stick m to stick n". Tasks must be performed in the order they appear on the stack (the request on the top of the stack is the most urgent one). We obtain the following program:

```
procedure move(i,m,n: integer);
begin
| make the stack empty
| put <i,m,n> into the stack
| {invariant: it remains to process
|             all the requests in the stack}
| while stack is not empty begin
|   | take the top of the stack into <j,p,q>
|   | if j = 1 then begin
|   |   | writeln ('move ', p, '->', q);
|   | end else begin
|   |   | s:=6-p-q; {s is the third stick; 1+2+3=6}
|   |   | put the triple <j-1,s,q> into the stack
|   |   | put the triple <1,p,q>   into the stack
|   |   | put the triple <j-1,p,s> into the stack
|   | end;
| end;
end;
```

(Please note that the triple put on the stack first will be the last request processed.) The stack of triples may be implemented as three separate stacks or one stack of records containing three integers (using a `record` type in Pascal).          □

**8.2.2.** (Communicated via A.K. Zvonkin by Andrzei Lissowski.) There are other non-recursive solutions of the Towers of Hanoi problem. Prove the correctness of the following solution: the *unused* stick (the stick that is neither the source nor the target of the move) should alternate cyclicly. (Another rule: alternately move the smallest ring and another ring, always moving the smallest one clockwise.)          □

**8.2.3.** In the recursive program that prints a decimal number (7.1.6), replace the recursion by a stack.

*Solution.* The digits are generated from right to left and put onto the stack. They are taken from the stack (in the reverse order) and printed.          □

**8.2.4.** Write a non-recursive program that prints all the vertices of a binary tree.

*Solution.* In this case, the stack of postponed tasks contains requests of two types: "print a vertex" and "print all the vertices of a subtree rooted at a given vertex". (We consider `nil` to be the root of an empty tree.) Therefore, the stack element is a pair ⟨request type, vertex number⟩.

When an element is taken off the stack, we either process it immediately (if it is a request of the first type) or put onto the stack the three requests caused by it (in one of six possible orderings).          □

**8.2.5.** What if we only want to count the number of vertices but not print them?

*Solution.* Instead of printing a vertex, we add 1 to a counter. In other words, the invariant is the following: (total number of vertices) = (counter) + (the total number of vertices in the subtrees whose roots are in the stack).          □

**8.2.6.** For some orderings (among six possible), the program that prints all vertices may be simplified. Show these simplifications.

*Solution.* If the order required is

$$\text{root, left subtree, right subtree,}$$

then a request to print the root may be processed immediately; thus we do not need to put it onto the stack.

A more complicated construction is necessary for the case

$$\text{left subtree, root, right subtree.}$$

In this case, all the requests in the stack (except the first one, which requests to print some subtree) are grouped into pairs

$$\text{print vertex x, print "right subtree" of x}$$

(that is, the subtree rooted at the right son of x). We can combine such pairs into requests of a special type and use an additional variable for the first request; in this way, all requests on the stack are homogeneous (have the same type).

For the symmetric case, similar simplifications are possible. Thus, for at least four of six possible orderings the program may be simplified.                    □

*Remark.* Another program that prints all the tree vertices is based on a program constructed in chapter 3. That program uses the command "down", which is not currently provided in the representation of trees. Therefore, we must keep a list of all vertices from the root to the current position (this list behaves like a stack).

**8.2.7.** Write a non-recursive version of Hoare's quicksort algorithm. How do we guarantee that the size of the stack does not exceed $C \log n$, where $n$ is the number of elements to be sorted?

*Solution.* The stack is filled with pairs $\langle i, j \rangle$, which are interpreted as requests to sort the corresponding intervals of the array. All such intervals are disjoint, therefore the size of the stack does not exceed $n$. To insure that the size of the stack is logarithmic, we follow the rule: "a larger request is pushed onto the stack first". Let $f(n)$ be the maximal size of the stack that may appear when sorting some array of length $n$ using this rule. We desire an upper bound for $f(n)$. Indeed, after the array is split into two fragments, the shorter one is sorted first (whereas the request to sort the longer one is kept on the stack); then the longer fragment is sorted. At the first stage, the size of the stack does not exceed $f(n/2) + 1$, and at the second stage it does not exceed $f(n - 1)$; therefore

$$f(n) \leqslant \max\{f(n/2) + 1, f(n - 1)\}$$

A simple induction argument gives $f(n) = O(\log n)$.                    □

## 8.3 Difficult cases

Finally, let us consider examples of recursion elimination not covered by the previous methods. Let $f$ be a function with nonnegative integer arguments and values defined by the equations

$$f(0) = a,$$
$$f(x) = h(x, f(l(x))) \quad (x > 0)$$

Here $a$ is some number while $h$ and $l$ are known functions. In other words, the value of $f$ at $x$ is determined by the value of $f$ at $l(x)$. We assume that for any $x$, the sequence

$$x, l(x), l(l(x)), \ldots$$

reaches 0. If we know in addition that $l(x) < x$ for all $x$, the computation of $f$ is trivial; just compute $f(0), f(1), f(2), \ldots$ sequentially.

**8.3.1.** Write a non-recursive program to compute $f$ in the general case.

*Solution.* To compute $f(x)$, compute the sequence

$$l(x), \; l(l(x)), \; l(l(l(x))), \ldots$$

until 0 appears. Now compute the values of $f$ for all terms of this sequence, going from right to left. □

The next example involves a more complicated case of recursion. (This example is hardly practical, and if it did appear in practice, it would probably be better to leave the recursion as is.)

Assume that a function $f$ with nonnegative integer arguments and values is defined by the equations

$$f(0) = a,$$
$$f(x) = h(x, f(l(x)), f(r(x))) \quad (x > 0),$$

where $a$ is a constant, and $l, r, h$ are (known) functions. We assume that if one starts from any nonnegative integer and applies functions $l$ and $r$ in some arbitrary order, one eventually gets 0.

**8.3.2.** Write a non-recursive program to compute $f$.

*Solution.* It is possible to construct a tree that has $x$ at the root, and has $l(i)$ and $r(i)$ as sons of vertex $i$ (unless $i$ is equal to 0, in which case it is a leaf). Then we may compute the values of $f$ from the leaves to the root. However, we'll use another approach.

By a *reverse Polish notation* (or *postfix notation*) we mean an expression where the function symbol is placed after all the arguments; parentheses are not used. Here are several examples:

| | |
|---|---|
| $f(2)$ | $2 \; f$ |
| $f(g(2))$ | $2 \; g \; f$ |
| $s(2, t(7))$ | $2 \; 7 \; t \; s$ |
| $s(2, u(2, s(5, 3)))$ | $2 \; 2 \; 5 \; 3 \; s \; u \; s$ |

Postfix notation allows us to compute the value of an expression easily using a so-called *stack calculator*. This calculator has a stack that we assume to be horizontal (the top of the stack is on the right), as well as number and function keys. When a number key is pressed, the number in question is put onto the stack. When a function key is pressed, the corresponding function is applied to the several arguments (according to its arity) taken from the stack. For example, if the stack contains the numbers

$$2 \; 3 \; 4 \; 5 \; 6$$

and the function key $s$ is pressed (we assume that $s$ is a function of two arguments), the new content of the stack is

$$2 \; 3 \; 4 \; s(5, 6)$$

Now let us return to our problem. The program employs a stack whose elements are nonnegative integers. It also uses a sequence of numbers and the symbols f, l, r, h (which we consider as a sequence of keys on a stack calculator). The invariant relation:

If the number stack represents the current state of a stack calculator and we press all the keys in the sequence, the stack contains only one number that is the required answer.

Suppose we want to compute $f(x)$. Put the number $x$ onto a stack and consider a sequence that contains only one symbol f. (The invariant relation is true.) Then the stack and the sequence are subjected to the following transformations:

| old stack | old sequence | new stack | new sequence |
|-----------|--------------|-----------|--------------|
| $X$ | $x\ P$ | $X\ x$ | $P$ |
| $X\ x$ | $\mathtt{l}\ P$ | $X\ l(x)$ | $P$ |
| $X\ x$ | $\mathtt{r}\ P$ | $X\ r(x)$ | $P$ |
| $X\ x\ y\ z$ | $\mathtt{h}\ P$ | $X\ h(x, y, z)$ | $P$ |
| $X\ 0$ | $\mathtt{f}\ P$ | $X\ a$ | $P$ |
| $X\ x$ | $\mathtt{f}\ P$ | $X$ | $x\ x\ \mathtt{l}\ \mathtt{f}\ x\ \mathtt{r}\ \mathtt{f}\ \mathtt{h}\ P$ |

Here $x$, $y$, $z$ are numbers, $X$ is a sequence of numbers, and $P$ is a sequence of numbers and the symbols f, l, r, h. In the last line, we assume that $x \neq 0$. This line corresponds to the equation

$$f(x) = h(x, f(l(x)), f(r(x)))$$

in postfix notation.

The transformations are performed until the sequence is empty. At that moment the invariant relation guarantees that the stack contains only one number, and this number is the answer required.

*Remark.* The sequence may be considered as a stack (whose top is on the left). □

# 9

# Graph algorithms

We have already seen several problems related to graphs. In this chapter we consider in more details two classes of problems: computation of shortest paths (section 9.1) and graph traversal algorithms (section 9.2).

## 9.1 Shortest paths

This section is devoted to different versions of one problem. Suppose a country has n cities numbered 1..n. For each pair of cities i and j, an integer a[i][j] is given that is the cost of a (nonstop) one-way plane ticket from i to j. We assume that flights exist between any two cities, and that a[k][k] = 0 for any k. In general, a[i][j] may be different from a[j][i]. Our goal is to find the minimal cost of a trip from one city (s) to another one (t) that takes into account all the possible travel plans (nonstop, one stop, two stops etc.). This minimal cost does not exceed a[s][t] but may be smaller. We allow a[i][j] to be negative for some i and j (you are paid if you agree to use some flight).

In the following problems, we compute the minimal cost for some pairs of cities, but first we have to check that our definition is correct.

**9.1.1.** Assume there is no cyclic travel plan with negative total cost. Prove that in this case a travel plan with minimal cost exists.

*Solution.* If a travel plan is long enough (includes more than n cities), it has a cycle, which may be omitted (because of our assumption). And there are finitely many travel plans involving n or fewer cities. □

Throughout the rest of this section, we assume that this condition (absence of negative cycles) is satisfied. (It is evident if all edge costs are nonnegative, but the latter condition is not always imposed.)

**9.1.2.** Find the minimal travel cost from the first city to all other cities in time $O(n^3)$.

A. Shen, *Algorithms and Programming*, Springer Undergraduate Texts in Mathematics and Technology, DOI 10.1007/978-1-4419-1748-5_9, © Springer Science+Business Media, LLC 2010

*Solution.* By MinCost(1,s,k) we denote the minimal travel cost from 1 to s with less than k stops. It is easy to see that MinCost(1, s, k+1) is equal to

$$\min\left\{\text{MinCost}(1,s,k), \min_{i=1..n} \{\text{MinCost}(1,i,k) + a[i][s]\}\right\}$$

The minimum on the right-hand side is taken over all possible places of the last stop before the final destination.

As we have seen in the solution of the preceding problem, the cycles can be eliminated, so the answer in question is MinCost(1,i,n) for all i = 1..n. We get the following program:

```
k:= 1;
for i := 1 to n do begin x[i] := a[1][i]; end;
{invariant: x[i] = MinCost(1,i,k)}
while k <> n do begin
  for s := 1 to n do begin
    y[s] := x[s];
    for i := 1 to n do begin
      if y[s] > x[i]+a[i][s] then begin
        y[s] := x[i]+a[i][s];
      end;
    end
    {y[s] = MinCost(1,s,k+1)}
  end;
  for i := 1 to n do begin x[s] := y[s]; end;
  k := k + 1;
end;
```

This algorithm is called the dynamic programming algorithm or Bellman–Ford algorithm.

**9.1.3.** Prove that the algorithm remains correct if the array y is not used; that is, if all changes are made in array x (just replace all y's by x's and delete redundant lines).

*Solution.* In this case the invariant is

$$\text{MinCost}(1,i,n) \leqslant x[i] \leqslant \text{MinCost}(1,i,k).$$

This algorithm may be improved in at least two ways. First, with the same running time $O(n^3)$, we can find the minimal travel cost i → j for *all* pairs i,j (not just i = 1). Second, we can compute all travel costs from a given vertex in time $O(n^2)$. (In the latter case, however, we require all flight costs a[i][j] to be nonnegative.)

**9.1.4.** Find the minimal travel costs i → j for all i,j in time $O(n^3)$.

*Solution.* For any k = 0..n consider the minimal travel cost from i to j *assuming intermediate stops are allowed only in cities* 1..k. This cost is denoted by A(i,j,k). Then

$$A(i,j,0) = a[i][j],$$

$$A(i,j,k+1) = \min\Big\{A(i,j,k),\ A(i,k+1,k) + A(k+1,j,k)\Big\}$$

(we either ignore city k+1 or use it as an intermediate stop; there is no reason to visit it twice). ☐

This algorithm is called the Floyd algorithm.

**9.1.5.** Find in $O(n^3)$ time whether a cyclic travel plan with negative total cost exists.

[*Hint.* Use Floyd's algorithm until the first negative cycle appears. Note that i = j should be allowed in A(i,j,k).] ☐

**9.1.6.** A table of cross rates for *n* currencies is given: rate[i][j] tells how many units of ith currency one get in exchange for one unit of jth currency. (Note that rate[i][j] could differ from 1/rate[j][i] due to transaction costs.) Check in $O(n^3)$ time whether an arbitrage deal is possible, i.e., you can make money by currency exchange alone.

[*Hint.* Taking logarithms converts money to distance.] ☐

**9.1.7.** Assume all costs a[i][j] are nonnegative. Find the minimal travel cost $1 \to i$ for all i = 1..n in time $O(n^2)$.

*Solution.* Our algorithm will *mark* cities during its operation. Initially, only city number 1 is marked. Finally, all cities are marked. For all the cities, a "current cost" is maintained. This cost is a number whose meaning is explained by the following invariant relation:

- for any marked city i, the current cost is the minimal cost of travel $1 \to i$; it is guaranteed that this minimal cost is obtained via a path through marked cities only;
- for any non-marked city i, the current cost is the minimal cost among all travel plans $1 \to i$ such that all intermediate stops are marked.

The set of marked cities is extended using the following observation: *for a non-marked city with minimal current cost* (*among all non-marked cities*), *the current cost is the true cost and is reached via a path going through marked cities only.*

Let us prove this. Assume that a shorter path exists. Consider the first non-marked city along this path: Even if we stop the trip in that city, the cost is already greater! (All costs are nonnegative.)

When a city is selected in this way, the algorithm marks it. To maintain the invariant, we update the current cost for non-marked cities. It is enough to take into account only those paths where the newly marked city is the last intermediate stop. This is easy to do since the minimal travel cost from the starting point to the newly marked city is already known.

If we store the set of marked cities in a Boolean array, we need $O(n)$ operations per city. ☐

This algorithm is called the Dijkstra algorithm.

**9.1.8.** There are n airports. For every i and j we know the baggage allowance on the flights from ith to jth airport. For a given starting point a and for all other airports x find the maximal weight that can be transported from a to x (using as many intermediate stops as necessary). The total time (for a given a and all x) should be $O(n^2)$.

[*Hint.* Replace sum by maximum in Dijkstra algorithm.]                                    □

The problem of finding the shortest path has a natural interpretation in terms of matrices. Assume that $A$ is the cost matrix for some carrier and $B$ is the cost matrix for another carrier. Suppose we want to make one stop along the way, using the first carrier (with matrix $A$) for the first flight and the second carrier ($B$) for the second flight. How much should we pay for the trip from i to j?

**9.1.9.** Prove that the costs mentioned above form a matrix that can be computed using a formula similar to the standard formula for matrix multiplication. The only difference is that the sum is replaced by a min-operation and the product is replaced by a sum:

$$C_{ij} = \min_{k=1,\ldots,n} \{A_{ik} + B_{kj}\}$$                                       □

**9.1.10.** Prove that matrix "multiplication" defined by the preceding formula is associative.                                                                              □

**9.1.11.** Prove that finding the shortest paths for all pairs of cities is equivalent to computation of $A^\infty$ for the cost matrix $A$ in the following sense. For the sequence $A, A^2, A^3, \ldots$ there exists an $N$ such that all elements $A^N, A^{N+1}$, etc. are equal to the matrix whose elements are minimal travel costs. (We assume that there are no cycles with negative cost.)                                                          □

**9.1.12.** How large should $N$ be in the preceding problem?                              □

The usual (unmodified) matrix multiplication may also be applied, but in a different situation. Let us assume that only some flights exist and let a[i][j] be equal to 1 if there is a (direct) flight from i to j; otherwise, a[i][j]=0. Compute the k-th power of the matrix a (in the usual sense) and consider its (i,j)-th element.

**9.1.13.** What is the meaning of this element?

*Solution.* It is the number of different travel plans from i to j using k flights (and k−1 intermediate stops).                                                              □

Let us return to our original problem (finding the shortest path). We can easily extend our algorithms to the case where not all pairs of cities are connected by direct flights. Indeed, we may assume that non-existing flights are infinitely expensive (or just very expensive), so our algorithms may be applied in this case too. However, a new question arises. The number of actual flights may be much smaller than $n^2$, so it is of interest to find algorithms that are more effective in this special case. First, we change the representation of the initial data: for each city we keep the number of outgoing flights and an array containing the destination points and costs.

**9.1.14.** Prove that the Dijkstra algorithm may be modified in such a way that if the number of cities is $n$ and the total number of flights is $m$, then no more that $C(n + m) \log n$ operations are required.

[*Hint.* What should we do at each step? We must choose a non-marked city with minimal current cost and update the data for all cities that can be reached by direct flight from this city. If there were an oracle to inform us which of the unmarked cities has minimal current cost, $C(n + m)$ operations would be enough. And an additional $\log n$-factor in the running time allows us to maintain the information needed to find the minimal value in the array (see the problem on p. 100).]                    □

## 9.2 Connected components, breadth- and depth-first search

The simplest possible case of the shortest path problem is when all the flight costs are 0 or $+\infty$. This means that we want to know whether it is possible to travel from $i$ to $j$, but do not worry about the price. In other words, we have a directed graph (a picture composed of points and arrows that connect some of the points) and we want to know which points are reachable from a given point via the arrows.

For this special case the algorithms given in the preceding section are not optimal. Indeed, a faster recursive program that solves this problem was given in chapter 7; its non-recursive version was shown in chapter 6. Now we add the following additional requirement: We not only want to list all the points (vertices) that are reachable from a given vertex via arrows (edges), but we also want to list them in a specific order. Two of the most popular instances of this are the so-called "breadth-first" and "depth-first" search.

### Breadth-first search

We are to list all the vertices of a directed graph that are reachable from a given vertex. The order is determined by the distance (minimal number of edges between a vertex and the given vertex); vertices at the same distance can be listed in any order.

**9.2.1.** Find an algorithm that performs breadth-first search in time $Cm$, where $m$ is the total number of outgoing edges of all reachable vertices.

*Solution.* This problem was considered in chapter 6, p. 100. Here we give a detailed solution. Let num[i] be the number of outgoing edges for vertex i, and let out[i][1],...,out[i][num[i]] be the endpoints of the edges emanating from vertex i. Here is the program (as it was written before):

```
procedure Print_Reachable (i: integer);
    {print all the vertices reachable from i,
        including the vertex i itself}
  var  X: subset of 1..n;
       P: subset of 1..n;
       q, v, w: 1..n;
       k: integer;
begin
  ...make X and P empty;
  writeln (i);
  ...add i to X, P;
  {(1) P = is a set of printed vertices; P contains i;
   (2) only vertices reachable from i are printed;
   (3) X is a subset of P;
   (4) all printed vertices which have an outgoing edge
       to a non-printed vertex, belong to X}
  while X is not empty do begin
    ...take some element of X into v;
    for k := 1 to num [v] do begin
      w := out [v][k];
      if w does not belong to P then begin
        writeln (w);
        add w to P;
        add w to X;
      end;
    end;
  end;
end;
```

If we do not worry about the order in which the reachable vertices are printed, it doesn't matter which element of X is chosen by the algorithm. Now we assume that X is a queue (first in, first out). In this case, the program prints all vertices reachable from i in order of increasing distance from i (distance is the number of edges on the shortest path from i). Let us prove this assertion.

By $V(k)$ we denote the set of vertices whose distance from i (in the sense described above) is $k$. The set $V(k + 1)$ is equal to the set

(endpoints of edges whose start points are in $V(k)$) $\setminus (V(0) \cup \ldots \cup V(k))$

Let us prove now that for a nonnegative integer $k = 0, 1, 2 \ldots$ there exists a point during the execution of the program (after one of the while-iterations) such that

- the queue contains all the elements of $V(k)$ and no other elements;
- all elements of $V(0), \ldots, V(k)$ and no others are printed.

For $k = 0$, it is the state before the first iteration. Now comes the induction step: Assume that at some point, the queue contains elements of $V(k)$. Those elements are processed one by one (the new elements are appended to the end of the queue and therefore cannot interfere). The endpoints of the edges emanating from the elements of $V(k)$ are printed and placed in the queue (unless they were printed earlier), exactly as in the equation for $V(k + 1)$ shown above. Therefore, when all elements of $V(k)$ are processed, the queue is filled with all the elements of $V(k + 1)$. $\qquad\square$

### Depth-first search

When thinking about depth-first search, it is convenient to represent a given graph as an image of a tree. Let us explain what we mean by this. Suppose some vertex $x$ of a directed graph is given. Assume that all vertices are reachable (via edges) from $x$. We construct a tree that may be called the "universal covering tree" of the graph. Its root is the point $x$, and it has the same outgoing edges as in the graph. The endpoints of those edges are sons of the root. Now consider any son $y$ of $x$ and all its outgoing edges. Their endpoints are the sons of $y$ in the tree. The difference between the graph and the tree is that different paths from $x$ to the same vertex of the graph now lead to different vertices of the tree. In other words, the vertex of the universal covering tree is a path in the graph starting from $x$. Its sons are paths that are one edge longer. Please note that the tree is infinite if the graph has (reachable) directed cycles.

There exists a natural mapping from the universal covering tree to the graph. For any vertex $y$ in the graph, the number of preimages is the number of paths from $x$ to $y$ in the graph. Therefore, if we visit the tree vertices in some order, we at the same time visit the vertices of the graph (but some graph vertices may be visited many times).

Assume that for any graph vertex the outgoing edges are numbered. Then for any vertex of the universal covering tree its sons are numbered. Let us visit tree vertices in the following order: first the root, then the subtrees rooted at the root's sons (in the given order of sons). An algorithm which traverses tree in that order was considered in chapter 7. This algorithm can be modified to traverse the graph avoiding visits to vertices already visited. Doing that, we get what is called "depth-first search".

Here is another description of depth-first search. Let us introduce a linear ordering on paths starting at a given vertex $x$. Any path precedes all its extensions. If two paths diverge at some vertex, they are ordered according to the ordering of the outgoing edges at that vertex. After that, vertices are ordered according to the minimal paths reaching them. This ordering is called *depth-first* ordering.

**9.2.2.** Write an algorithm for depth-first search.

[*Hint.* Take a program that traverses a tree (root $\rightarrow$ left subtree $\rightarrow$ right subtree) from chapters 7 or 8 and modify it. The main difference is that we do not want to revisit any visited vertex. Therefore, if we are at an already-visited vertex, we do nothing. (If a path is not minimal among all paths going to some vertex, all its extensions are not minimal as well, and can be safely ignored.)] $\qquad\square$

*Remark.* Recall that in chapter 8 two possible non-recursive algorithms for tree traversal were mentioned (p. 126). Both versions may be used for depth-first search.

Depth-first search is used in several graph algorithms (sometimes in a modified form).

**9.2.3.** An undirected graph is called a *bipartite graph* if its vertices may be colored in two colors in such a way that each edge connects vertices of different colors. Find an algorithm that checks whether a graph is a bipartite graph in time $C \cdot$ (number of edges + number of vertices).

[*Hint*. (a) Each connected component may be considered separately. (b) After we choose the color of some vertex, the colors of all other vertices of the same component are uniquely determined.]                                                   $\square$

*Remark*. In this problem we may use breadth-first as well as depth-first search.

**9.2.4.** Write a non-recursive algorithm for topological sorting of a directed graph without cycles. (For a recursive algorithm, see p. 113.)

*Solution*. Assume that the graph has vertices 1..n. For every vertex i, we know the number num[i] of outgoing edges. The endpoints of these outgoing edges are dest[i][1], ..., dest[i][num[i]]. We adopt the following terminology: the outgoing edges are listed "from left to right" (so dest[i][1] is "on the left" of dest[i][2], etc.).

Our goal is to print all the vertices of the graph; the requirement is that the endpoint of any edge is printed before its starting point. We assume that the graph has no cycles (otherwise this is impossible).

Let us add to the graph an auxiliary vertex 0 that has n outgoing edges to 1, ..., n. If it is printed and the requirement is fulfilled, then all other vertices are already printed.

Our algorithm maintains a path that starts at 0 (the auxiliary vertex) and traverses the graph edges. The length of this path is kept in an integer variable m. The path is formed by the vertices vert[1]..vert[m] and edges having numbers edge[1]..edge[m]. The number edge[s] refers to the numbering of all outgoing edges of the vertex vert[s]. Therefore, for all s, the following inequality holds:

$$\text{edge[s]} \leqslant \text{num[vert[s]]}$$

as well as the equality

$$\text{vert[s+1]} = \text{dest[vert[s]][edge[s]]}.$$

Note that the endpoint of the last edge in the path (i.e., dest[vert[m]][edge[m]]) is not included in the array vert. Moreover, we make an exception for the last edge and allow it to point "nowhere": edge[m] may be equal to num[vert[m]]+1.

The algorithm prints the vertices of the graph; a vertex is printed only after all the vertices where the outgoing edges go are printed. Moreover, the following requirement (I) is fulfilled:

all vertices in the path, except the last one (i.e., the vertices
vert[1]..vert[m]), are not printed, but if we leave our path turning
to the left, we immediately come to an already printed vertex.

Here is the algorithm in full:

```
m:=1; vert[1]:=0; edge[1]:=1;
{(I)}
while not((m=1) and (edge[1]=n+1)) do begin
  if edge[m]=num[vert[m]]+1 then begin
    {path leads to nowhere, therefore all vertices
        following vert[m] are printed and we may
        print vert[m]}
    writeln (vert[m]);
    m:=m-1; edge[m]:=edge[m]+1;
  end else begin
    {edge[m] <= num[vert[m]], path ends in a real
        vertex}
    lastvert:= dest[vert[m]][edge[m]];
    if lastvert is printed then begin
      edge[m]:=edge[m]+1;
    end else begin
      m:=m+1; vert[m]:=lastvert; edge[m]:=1;
    end;
  end;
end;
{the path immediately goes to nowhere, so all the
vertices on the left (1..n) are printed}
```
□

**9.2.5.** Prove that if the graph has no cycles, this algorithm terminates.

*Solution.* Assume that this is not true. Any vertex may be printed at most once, so the vertices are not printed after some point. In a graph without cycles, the path length is limited (no vertex can appear in a path twice); therefore, after some point the path never becomes longer. After that, the only possibility is an increase in edge [m], but this cannot happen infinitely many times. □

**9.2.6.** Prove that the running time of the previous algorithm is $O$(number of vertices + number of edges). □

**9.2.7.** Modify the algorithm in such a way that it can be applied to any graph. The algorithm should either find a cycle (if it exists) or perform a topological sort (if there are no cycles). □

# 10

# Pattern matching

Pattern matching is a basic operation for string processing. Given two strings, we want to find out whether one of them is a substring of the other one. We start in section 10.1 with a simple example. Then in section 10.2 we explain what kind of difficulties we encounter if the pattern contains repetitions. After some preparations in section 10.3 (simple lemmas about prefixes and suffixes) we consider a classical linear time algorithm for pattern matching called the Knuth–Morris–Pratt algorithm (section 10.4). Some other well-known algorithms are considered in the next two sections: in section 10.5 we consider a simplified version of the Boyer–Moore algorithm that can be very efficient for large alphabets. In section 10.6 we consider the Rabin–Karp randomized algorithm. In section 10.7 we discuss an important notion of finite automaton (cf. chapter 5) in its full generality and establish its connection with regular expressions. Finally, in section 10.8 we consider a linear-time pattern matching algorithm that first processes a string and then gets the pattern that should be found in that string.

## 10.1 Simple example

**10.1.1.** The character string $x[1]..x[n]$ is given. Check if it contains the substring abcd.

*Solution.* There are approximately n (or n-3, to be exact) positions where a substring of length 4 may be found. For each position, we can check whether the substring appears in that position. This would require approximately $4n$ comparisons.

However, there is a more efficient approach. Reading the string $x[1]..x[n]$ from left to right, we are looking for the character a. After it appears, we look for the character b (immediately after a), then for c, and finally d. If our expectations are met, the substring abcd is found. If one of the letters is not found where expected, we start from scratch looking for a again.

This simple algorithm can be described in different terms. In the framework of so-called *finite automata*, we say that while scanning x from left to right the algorithm

A. Shen, *Algorithms and Programming*, Springer Undergraduate Texts
in Mathematics and Technology, DOI 10.1007/978-1-4419-1748-5_10,
© Springer Science+Business Media, LLC 2010

is in one of the following "states": the initial state (0), the state "immediately after a" (1), "immediately after ab" (2), "immediately after abc" (3) and "immediately after abcd" (4). When reading the next character, we change the state according to the following rule:

| Current state | Next character | New state |
|:---:|:---:|:---:|
| 0 | a | 1 |
| 0 | except a | 0 |
| 1 | b | 2 |
| 1 | a | 1 |
| 1 | except a,b | 0 |
| 2 | c | 3 |
| 2 | a | 1 |
| 2 | except a,c | 0 |
| 3 | d | 4 |
| 3 | a | 1 |
| 3 | except a,d | 0 |

As soon as we come to state 4, or the input string is exhausted, the search is complete.

This process can be illustrated by a diagram: a token moves from one circle to another along the edge that is labeled with current input letter. To make this process well defined, each vertex should have outgoing edges labeled by all the input letters (one for each letter).

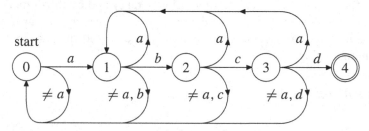

The corresponding program is straightforward (we indicate the new state even if it coincides with the old one, but those lines may be omitted):

```
i:=1; state:=0;
{i is the index of the first unread character;
 state is the current state}
while (i <> n+1) and (state <> 4) do begin
  if state = 0 then begin
    if x[i] = a then begin
    | state:= 1;
    end else begin
    | state:= 0;
    end;
  end else if state = 1 then begin
```

```
    if x[i] = b then begin
    | state:= 2;
    end else if x[i] = a then begin
    | state:= 1;
    end else begin
    | state:= 0;
    end;
  end else if state = 2 then begin
    if x[i] = c then begin
    | state:= 3;
    end else if x[i] = a then begin
    | state:= 1;
    end else begin
    | state:= 0;
    end;
  end else if state = 3 then begin
    if x[i] = d then begin
    | state:= 4;
    end else if x[i] = a then begin
    | state:= 1;
    end else begin
    | state:= 0;
    end;
  end;
end;
answer := (state = 4);
```

In other words, at any point we keep information about the *maximal prefix of the pattern abcd which is a suffix of the substring already read.* (Its length is the value of the variable state.)                                                                                    □

Let us recall the terminology used. A *string* is an arbitrary finite sequence of elements of a set called an *alphabet*; its elements are called *letters*. If we discard some letters at the end of a string, we get another string, which is called a *prefix* of the first string. Any string is a prefix of itself. The *suffix* of a string is what remains after several initial letters of a string are discarded. Every string is a suffix of itself. A *substring* is obtained when we discard some letters both at the beginning and the end of a string. (In other words, substrings are prefixes of suffixes, as well as suffixes of prefixes.)

In terms of "inductive functions" (see section 1.3) we can describe the situation as follows: Consider a function whose arguments are strings and whose values are either "True" or "False". The function has value "True" for all strings containing substring abcd. This function is not inductive, but it does have an inductive extension:

$x \mapsto$ the length of the maximal prefix of abcd that is a suffix of $x$

## 10.2 Repetitions in the pattern

**10.2.1.** Can the previous algorithm be used for any other string instead of abcd?

*Solution.* A problem arises when the pattern contains repetitions. For example, suppose we are looking for substring ababc. Assume that a appears, then b, a, and b again. At this point, we are eagerly waiting for c. If the letter d appears instead, we should start from scratch. However, if the letter a appears, we still have a chance that b and c follow and the pattern is found.

Here is an illustration:

```
x y z a b a b a b c ...   ← input string
       a b a b c          ← pattern was expected here
           a b a b c      ← but it is here
```

In other words, at the point

```
x y z a b a b|            ← input string
     a b a b|c            ← pattern was expected here
     a b|a b c            ← but it is here
```

there are two possible pattern positions to be tested.                               □

However, a finite automaton that reads the input string letter-by-letter, changes its state according to some table, and says (after the input string is exhausted) whether the input string contains the given substring, is still possible.

**10.2.2.** Construct such an automaton. Show all its states and the transition table (which determines a new state as a function of an old state and an input character).

*Solution.* As before, the current state is the length of the maximal prefix of the pattern that is also a suffix of the currently read part of the string. There are six states: 0, 1 (a), 2 (ab), 3 (aba), 4 (abab), 5 (ababc). The transition table is as follows:

| Current state | Next character | New state |
|---|---|---|
| 0 | a | 1 (a) |
| 0 | except a | 0 |
| 1 (a) | b | 2 (ab) |
| 1 (a) | a | 1 (a) |
| 1 (a) | except a,b | 0 |
| 2 (ab) | a | 3 (aba) |
| 2 (ab) | except a | 0 |
| 3 (aba) | b | 4 (abab) |
| 3 (aba) | a | 1 (a) |
| 3 (aba) | except a,b | 0 |
| 4 (abab) | c | 5 (ababc) |
| 4 (abab) | a | 3 (aba) |
| 4 (abab) | except a,c | 0 |

Consider the second (from below) line in this table as an example. If the processed part is ended by abab and the next letter is a, the new processed part is ended by ababa. The maximal prefix of ababc, which is also a suffix of the processed part, is aba. □

*Question*: As we said, the difficulty appears because there are several possible positions of the pattern; each position corresponds to some prefix of the pattern that is also a suffix of the input string. The finite automaton remembers only the longest one. What about others?

*Answer*. The longest prefix-suffix $X$ determines all other prefix-suffixes. Namely, prefix-suffixes of the processed part are prefixes of $X$ that are also suffixes of $X$.

It is easy to write a transition table and a program for any fixed pattern. However, we want to write a general program that will search for any given pattern in any given input string. The following approach may be used. Consider a program that has two stages. In the first stage, it examines the pattern and constructs a transition table for that pattern. In the second stage, it reads the input string and behaves according to the transition table. Such an approach is often used for more complicated patterns (see section 10.7 below), but for a substring search there is a more direct and efficient method called the Knuth–Morris–Pratt algorithm. (A similar approach was suggested by Yu. Matijasevich.) We start with some auxiliary lemmas.

## 10.3 Auxiliary lemmas

For any string $X$, consider all the prefixes of $X$ that are at the same time suffixes of $X$. Choose the longest one (not counting $X$ itself), which is denoted by $l(X)$.

For example, $l(\text{aba}) = \text{a}$, $l(\text{abab}) = \text{ab}$, $l(\text{ababa}) = \text{aba}$, $l(\text{abc}) = $ the empty string.

**10.3.1.** Prove that all strings $l(X), l(l(X)), l(l(l(X)))$, etc. are prefixes of $X$.

*Solution*. Each of them is a prefix of the preceding one. (And any prefix of a prefix of $X$ is also a prefix of $X$.) □

For the same reason all such strings are suffixes of $X$ as well.

**10.3.2.** Prove that the sequence in the preceding problem is finite (the last string is empty).

*Solution*. Each subsequent string is shorter than the preceding one (since $l(Y)$ is shorter than $Y$ for any $Y$). □

**10.3.3.** Prove that any string that is both a prefix and a suffix of $X$ (except for $X$ itself) is listed in the sequence $l(X), l(l(X)), \ldots$.

*Solution*. Let $Y$ be both a prefix of $X$ and a suffix of $X$. The string $l(X)$ is the longest string having this property, so $Y$ is not longer than $l(X)$. Both $Y$ and $l(X)$ are prefixes of $X$, and the shorter one is a prefix of the longer one. Thus $Y$ is a prefix of $l(X)$. For the same reason, $Y$ is a suffix of $l(X)$. Using an induction argument, we assume that the statement in question is true for all strings shorter than $X$. In particular, it is true for $l(X)$. So the string $Y$, being a prefix and a suffix of the string $l(X)$, is either equal to $l(X)$ or one of the strings $l(l(X)), l(l(l(X))), \ldots$.     □

## 10.4 Knuth–Morris–Pratt algorithm

The Knuth–Morris–Pratt (KMP) algorithm takes a string

$$X = \texttt{x[1]x[2]..x[n]}$$

as input and scans it from left to right. The output is the sequence of nonnegative integers $\texttt{L[1]} \ldots \texttt{L[n]}$ such that

$$\texttt{L[i]} = \text{the length of } l(\texttt{x[1]..x[i]})$$

(the function $l$ is defined in the preceding section). In other words, $\texttt{L[i]}$ is the length of the maximal prefix of $\texttt{x[1]..x[i]}$ that is simultaneously a suffix of $\texttt{x[1]..x[i]}$.

**10.4.1.** How can we use the KMP algorithm to check whether a given string $\texttt{A}$ is a substring of a string $\texttt{B}$?

*Solution.* Apply the KMP algorithm to the string $\texttt{A\#B}$, where $\texttt{\#}$ is a special character that does not appear in $\texttt{A}$ or $\texttt{B}$. The string $\texttt{A}$ is a substring of $\texttt{B}$ if and only if the array $\texttt{L}$ (which is the output of the KMP algorithm) contains a number equal to the length of $\texttt{A}$.     □

**10.4.2.** How do we fill the table $\texttt{L[1]..L[n]}$?

*Solution.* Assume that the first $\texttt{i}$ values $\texttt{L[1]..L[i]}$ are already known. We read the next input character (i.e., $\texttt{x[i+1]}$) and compute $\texttt{L[i+1]}$.

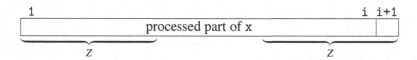

How do we find $\texttt{L[i+1]}$? It is the length of the longest prefix $Z$ of the string $\texttt{x[1]..x[i+1]}$ that is at the same time a suffix of this string. Any string $Z$ having this property (except for the empty string) is obtained from some string $Z'$ by adding the letter $\texttt{x[i+1]}$. The string $Z'$ is both a prefix and a suffix of the string $\texttt{x[1]..x[i]}$. However, it is not the only requirement for $Z'$; another requirement is that $Z'$ is followed (as a prefix of $\texttt{x[1]..x[i]}$) by $\texttt{x[i+1]}$.

Therefore, the string $Z$ may be found as follows. Consider all the prefixes $Z'$ of the string $\texttt{x[1]}\ldots\texttt{x[i]}$ that are also the suffixes of this string. Then choose the longest one that is followed (as a prefix of $\texttt{x[1]}\ldots\texttt{x[i]}$) by $\texttt{x[i+1]}$. Adding $\texttt{x[i+1]}$ produces the string $Z$.

Now it is time to use the lemmas proved earlier. Recall that all strings that are both prefixes and suffixes may be obtained by applying the function $l$ iteratively. Here is the program:

```
     i:=1; L[1]:= 0;
     {the table L[1]..L[i] is filled correctly}
     while i <> n do begin
       len := L[i]
       {len is the length of a prefix of x[1]..x[i] that is
          its suffix; all longer prefixes-suffixes were
          tested without success}
       while (x[len+1] <> x[i+1]) and (len > 0) do begin
         {this prefix does not fit also, we should apply L}
         len := L[len];
       end;
       {we either have found the longest prefix that
        fits our requirements (and its length is len)
        or have found that it does not exist (len=0)}
       if x[len+1] = x[i+1] do begin
         {x[1]..x[len] is the longest prefix that fits}
         L[i+1] := len+1;
       end else begin
         {there are no good prefixes}
         L[i+1] := 0;
       end;
       i := i+1;
     end;
   end;                                                              □
```

**10.4.3.** Prove that the number of operations in the above algorithm is limited by $Cn$ for some constant $C$.

*Solution.* This is not obvious, because one input character may cause many iterations in the inner loop. However, each iteration in the inner loop decreases len by at least 1, so in this case, L[i+1] will be significantly smaller than L[i]. On the other hand, while i is increased by 1, the value of L[i] may increase by at most 1, therefore the values of i that require many iterations in the inner loop are rare.

Formally, we use the inequality

$$L[i+1] \leqslant L[i] - (\text{the number of iterations at step i}) + 1$$

or

$$(\text{the number of iterations at step i}) \leqslant L[i] - L[i+1] + 1$$

Summing these inequalities over i, we get the required upper bound for the total number of iterations.                                                    □

**10.4.4.** Imagine that we use this algorithm to determine whether a string X of length n is a substring of a string Y of length m. (We explained above how to do that using a "separator" #.) The algorithm runs in time $O(n + m)$ and uses memory of size $O(n + m)$. Find a way to do this using memory of size $O(n)$ (which may be significantly less if the pattern is short and the string is long).

*Solution.* Start applying the KMP algorithm to the string A#B. Wait until the algorithm computes all the values L[1], ..., L[n] for the word X of length n. All those values are stored. From then on, we keep only the value L[i] for the current i; we only need L[i] and the table L[1]..L[n] to compute L[i+1].                     □

In practice, the words X and Y are usually separated, so the scan of X and the scan of Y should be implemented as two different loops. (This also makes the separator # unnecessary.)

**10.4.5.** Write the program discussed in the last paragraph: It checks whether a string X = x[1]..x[n] is a substring of a string Y = y[1]..y[m].

*Solution.* First we fill the table L[1]...L[n] as before. Then we execute the following program:

```
j:=0; len:=0;
{len is the length of a longest prefix of X which is
  a suffix of y[1]..y[j]}
while (len <> n) and (j <> m) do begin
    while (x[len+1] <> y[j+1]) and (len > 0) do begin
        {this prefix does not fit}
        len := L[len];
    end;
    {we have found the prefix that fits or
        have found that it does not exist}
    if x[len+1] = y[j+1] do begin
        {x[1]..x[len] is the longest prefix that fits}
        len := len+1;
    end else begin
        {no prefixes fit}
        len := 0;
    end;
    j := j+1;
end;
{if len=n, X is a substring of Y;
    otherwise we reached the end of Y not finding X}        □
```

## 10.5 Boyer–Moore algorithm

This algorithm attains a goal that seems impossible at first: In a typical situation, it reads only a tiny fraction of all the characters of a string in which the pattern is searched. How can this be done? The idea is rather simple. Suppose we are searching for the pattern abcd in a string $X$. Check the fourth character of $X$. If it is, say, e, there is no need to look at the first three characters, because our pattern does not contain e and may start only after the fourth position.

We show below a simplified version of the Boyer–Moore algorithm that does not guarantee good running time in all cases.

Let X = x[1]..x[n] be the pattern we are searching for. For any character s, we find the rightmost occurrence of s in the string X; that is, the maximal k such that x[k] = s. This information is stored in an array pos[s]. If the character s does not appear in the pattern at all, it is convenient to put pos[s] := 0 (see below).

**10.5.1.** How do we fill the array pos?

*Solution.*

```
...let all pos[s] be equal to 0
for i:=1 to n do begin
| pos[x[i]]:=i;
end;                                                          □
```

The program searches for the pattern x[1]..x[n] in the string y[1]..y[m]. When searching, store in the variable last the index of the input character that is aligned with the last character of the pattern (in the current pattern position). Initially, last = n (the length of the pattern); then last increases gradually.

```
last:=n;
{all previous positions of the pattern are checked}
while last <= m do begin {the work is not finished}
  if x[n] <> y[last] then begin
    {the last characters differ}
    last := last + (n - pos[y[last]]);
    {n - pos[y[last]] is the minimal shift of the
        pattern that makes the character y[last]
        match the corresponding character in the
        pattern. If y[last] does not appear in the
        pattern, the new pattern position starts
        immediately after y[last]}
  end else begin
    {x[n] = y[last]}
    check if the current position is okay; that is,
    if x[1]..x[n] = y[last-n+1]..y[last].
    If yes, inform about that.
    last := last+1;
  end;
end;
```

It is recommended to check the condition x[1]..x[n] = y[last-n+1]..y[last] from right to left starting from the last position (where the coincidence is already tested). We also obtain a small optimization if we store n-pos[s] instead of pos[s] (avoiding subtraction at each step); n-pos[s] is the number of characters to the right of the rightmost occurrence of character s in the pattern.

Different versions of this algorithm exist. For example, we may replace the line last:=last+1 by last:=last+(n-u), where u is the position of the second (from the right) occurrence of the character x[n] in the pattern.

**10.5.2.** What modifications in the program are necessary?

*Solution.* To fill up the table pos, we use the line

```
for i:=1 to n-1 do...
```

(all other lines remain the same); in the main program we replace `last:=last+1` by

```
last:= last+n-pos[y[last]];
```
                                                                                     □

We have described a simplified version of the Boyer–Moore algorithm sometimes require significantly more that n operations (mn in the worst case), so the worst-case behavior of the Knuth–Morris–Pratt algorithm is much better.

**10.5.3.** Give an example where a pattern of length n is not a substring of a given string of length m, but the program above requires mn operations to determine this.

*Solution.* Assume that the pattern is baaa..aa and the string contains n letters a. Then at each step we need n comparisons to discover that the pattern is not a substring.                                                                              □

The complete (not simplified) Boyer–Moore algorithm guarantees that the number of operations does not exceed $C(m + n)$ in the worst case. It uses ideas similar to those in the KMP algorithm. Suppose we compare the pattern and the string from right to left. Assume that we find the coincident suffix $Z$ of the pattern, but the characters before $Z$ in the input string and in the pattern are different. What do we know about the input string at that point? We have found a fragment equal to $Z$ that is preceded by a character that differs from the character in the pattern. This information may allow us to shift the pattern to the right several positions. These shifts should be computed in advance for all suffixes $Z$ of the pattern. One can prove that all operations (the computation and use of the shift table) can be performed in time $C(m + n)$.

## 10.6 Rabin–Karp algorithm

This algorithm is also based on a simple idea. Suppose we are looking for a pattern of length n in a string of length m. Let us make a sliding window and move it along the input string. Our goal is to check whether the substring in the window coincides with the given pattern.

We want to avoid character-by-character comparison and find a faster method. Let us consider some function defined on strings of length $n$. If this function takes on different values when applied to both the pattern and the substring in the window, we may be sure that there is no match. Only if the function values coincide, we have to compare strings character-by-character.

What do we gain? It seems that we have achieved nothing because to compute the function value for the substring in the window, we have to read all the characters in the window anyway. So why not just compare them with the pattern characters? Some gain, however, is still possible for the following reason. When we shift the window, the substring in it does not change completely; a single character is appended on the right and deleted on the left. If our function is well chosen, we may compute its new value quickly, knowing its old value and the added/deleted characters.

**10.6.1.** Find an example of such a "well chosen" function.

*Solution.* Replace all characters in the pattern by their codes, which are assumed to be integers. The sum of all codes is such a function. (Indeed, after the shift, we only have to add the numeric value of the new character and subtract the numeric value of the old character.) □

Given any function, most likely there are distinct strings that are mapped to the same value. For the same pair of strings another function may indeed produce distinct values. So let us have a pool of functions and begin the algorithm by choosing one of the functions at random. Then an adversary who wants to choose the worst problem instance will not know which function it is working against.

**10.6.2.** Give an example of a family of easily computable functions (in the sense explained above). □

*Solution.* Let us choose some number $p$ (presumably prime; see below) and some residue $x$ modulo $p$. Each string of length $n$ is considered as a sequence of integers (characters are replaced by their numeric codes). Those integers are taken to be coefficients of a polynomial of degree $n - 1$. We compute the value of this polynomial modulo $p$ at the point $x$. This construction provides one function of the family (for each $p$ and $x$ we get another function). When the window is shifted by 1, we subtract the term of the highest degree ($x^{n-1}$ should be computed in advance), multiply by $x$, and add the constant term.

The following arguments show that the coincidence of function values (for different arguments) is not very likely. Assume that $p$ is fixed and is prime. Let $X$ and $Y$ be two different words of length $n$. Then the corresponding polynomials are different. (We assume that different characters have different codes modulo $p$, so we need $p$ to be larger than the size of the alphabet.) The coincidence of function values on $X$ and $Y$ means that two different polynomials coincide at $x$; that is, $x$ is a root of their difference. This difference is a nonzero polynomial of degree $n - 1$ and can have at most $n - 1$ roots. Therefore, if $n$ is much smaller than $p$, the chances for the random $x$ to be a root are negligible.

## 10.7 Automata and more complicated patterns

Rather than a specific string, we may search for a string of some type. For example, we may look for a substring of type a?b where ? denotes any single character. In other words, we are looking for characters a and b with exactly one character in between.

**10.7.1.** Construct a finite automaton that checks if the pattern a?b is present in the input string.

*Solution.* While reading the input string, the automaton keeps track of whether the character a is present at the two last positions. The automaton has states 00, 01, 10, 11 with the following meanings:

00 no a in the last two positions
01 a is in the last position but not in the position immediately
   before it
10 a is in the position before the last one but not in the last
   position
11 the processed part of the input string ends with aa

Here is the transition table:

| Current state | Next character | New state |
|---|---|---|
| 00 | a | 01 |
| 00 | not a | 00 |
| 01 | a | 11 |
| 01 | not a | 10 |
| 10 | a | 01 |
| 10 | b | found |
| 10 | not a and not b | 00 |
| 11 | a | 11 |
| 11 | b | found |
| 11 | not a and not b | 10 |

□

Another widely used notation in a pattern is an asterisk (∗), which is matched by any string (including the empty string). For example, the pattern ab∗cd means that we are looking for any occurrence of ab followed by cd (the distance between ab and cd is arbitrary).

**10.7.2.** Construct a finite automaton that checks if the input string contains the pattern ab∗cd (in the sense just described).

*Solution.*

| Current state | Next character | New state |
|---|---|---|
| initial | a | a |
| initial | not a | initial |
| a | b | ab |
| a | a | a |
| a | not a and not b | initial |
| ab | c | abc |
| ab | not c | ab |
| abc | d | found |
| abc | c | abc |
| abc | not c and not d | ab |

□

Another type of search occurs when we are looking for a substring that belongs to a given finite set of strings.

**10.7.3.** Assume that strings $X_1, \ldots, X_k$ (patterns) and a string $Y$ are given. Check if one of the strings $X_i$ is a substring of the string $Y$. The number of operations should not exceed the total length of all the strings ($X_i$ and $Y$) multiplied by some constant which does not depend on $k$.

*Solution.* The obvious approach is to check all the $X_i$ separately (using one of the algorithms given above). However, this method does not satisfy the speed requirements (since we have to read the string $Y$ many, in fact, $k$ times).

Let us look at another aspect of the problem. For each pattern $X_i$, there exists a finite automaton that tests for the presence of $X_i$. These automata may be combined into one automaton whose set of states is the product of the sets of states for all the automata. This set is very large. However, most of its elements are unreachable and may be discarded.

This idea is used below (in a modified form).

Let us recall the Knuth–Morris–Pratt algorithm. While reading the input string, the KMP algorithm keeps the maximal prefix of the pattern that is a suffix of the processed part of the input string. Now we need to keep this information (the longest prefix that is a suffix of the processed part) for all the patterns. The crucial remark is: It is enough to keep the longest one, because all others are uniquely determined by the longest one. Indeed, let $X$ be the longest prefix of some pattern that is a suffix of the processed part of the input string. Then for any pattern $P$, the longest prefix of $P$ being a suffix of the processed part is the longest prefix of $P$ being a suffix of $X$.

All the patterns may be "glued" together to form a tree if we "splice" together equal prefixes. For example, the set of patterns

$$\{aaa, aab, abab\}$$

corresponds to the tree

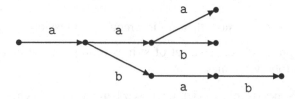

Here is the formal definition: any prefix of any pattern is a tree vertex; a father of a vertex is obtained by deleting the last character.

While reading the input string, we traverse this tree. The current position is the maximal (rightmost) vertex that is a suffix of the processed part of the input string (that is, the longest suffix of the processed part being a prefix of one of the patterns).

Let us introduce a function $l$ whose arguments and values are tree vertices, namely, $l(P)$ = maximal tree vertex that is a (proper) suffix of $P$. (Recall that tree vertices are strings.) The following result will be used:

**10.7.4.** Let $P$ be a tree vertex. Prove that the set of all tree vertices that are (proper) suffixes of $P$ is $\{l(P), l(l(P)), \ldots\}$

*Solution.* See the proof of the similar assertion for the Knuth–Morris–Pratt algorithm.                                                                            □

Now it is clear what the algorithm (or automaton) should do if it is at the vertex $P$ and the next input character is z: It should consider sequentially the vertices $P, l(P)$, $l(l(P)), \ldots$ until it finds the vertex that has an outgoing (to the right) edge labeled "z". The endpoint of that edge is the next position of the algorithm (next state of the automaton).

It remains to show how to compute the values of the function $l$ for all tree vertices. This is done as before using the values of $l$ for shorter strings to compute the next value of $l$. Therefore, we should consider all tree vertices in order of increasing length. It is easy to see that this can be done in the required time. (Please note that the constant in the upper bound for the running time depends on the cardinality of the alphabet.) For a discussion of the methods used to store the tree, see chapter 9. □

The general question arises: Which properties of strings can be tested using finite automata? It turns out that there is an easily defined class of patterns that correspond to finite automata. These patterns are called "regular expressions".

**Definition.** Let $\Gamma$ be a finite alphabet. We assume that $\Gamma$ does not contain six symbols $\Lambda$, $\varepsilon$, $($, $)$, $*$ and $|$ (these symbols will be used for constructing regular expressions; therefore, we should not mix them with letters from $\Gamma$). *Regular expressions* are constructed according to the following rules:

(a) any letter from $\Gamma$ is a regular expression;
(b) the symbols $\Lambda$, $\varepsilon$ are regular expressions;
(c) if $A, B, C, \ldots, E$ are regular expressions, then $(ABC \ldots E)$ is a regular expression;
(d) if $A, B, C, \ldots, E$ are regular expressions, then $(A | B | C | \ldots | E)$ is a regular expression;
(e) if $A$ is a regular expression, then $A*$ is a regular expression.

Each regular expression defines a set of strings (composed of characters from $\Gamma$) according to the following rules:

(a) A letter corresponds to a singleton whose element is a one-character string containing this letter;
(b) The symbol $\varepsilon$ corresponds to the empty set; the symbol $\Lambda$ corresponds to the singleton whose element is the empty string;
(c) the regular expression $(ABC \ldots E)$ corresponds to the set of all strings obtained as follows: take a string from the set that corresponds to $A$, a string from the set that corresponds to $B$, to $C, \ldots$, and to $E$ and concatenate all those strings in the given order (*concatenation* of sets);

(d) the regular expression $(A \mid B \mid C \mid \ldots \mid E)$ corresponds to the union of the sets that correspond to expressions $A, B, C, \ldots, E$;

(e) the regular expression $A*$ corresponds to the *iteration* of a set corresponding to $A$; that is, to the set of all strings that may be cut into pieces in such a way that each piece belongs to the set corresponding to $A$. (In particular, the set corresponding to $A*$ always contains the empty string.)

Sets that correspond to regular expressions are called *regular* sets. Here are several examples:

| Expression | Set |
|---|---|
| (a\|b)* | All strings composed of a and b |
| (aa)* | All strings of even length composed of as, including the empty string |
| (Λ\|a\|b\|aa\|ab\|ba\|bb) | all strings of length at most 2 composed of a and b |

**10.7.5.** Find a regular expression corresponding to the set of all strings composed of a and b that contain an even number of as.

*Solution.* The expression b* defines the set of all strings without a; the expression (b* a b* a b*) defines the set of all strings with exactly two as. It remains to take the union of these two sets and then to apply iteration:

$$( (b* a b* a b*) \mid b* )*$$

Another possible answer:

$$( (b* a b* a)* b* )$$  □

**10.7.6.** Write a regular expression that defines a set of strings composed of a, b, c having bac as a substring.

*Solution.* ((a\|b\|c)* bac (a\|b\|c)*)  □

**10.7.7.** Prove that there exists a regular expression that defines a complement of this set, i.e., the set of strings composed of a, b, c that do not contain bac as a substring.

[*Hint.* This is a more difficult task. Probably the simplest solution is to use the equivalence between regular expressions and finite automata (cf. Problem 10.7.14).] □

Now the general pattern-matching problem may be stated as follows: check whether an input string belongs to the set corresponding to a given regular expression.

**10.7.8.** What regular expressions are equivalent to the patterns a?b and ab*cd used as examples earlier? (Please note that the symbol * in the pattern ab*cd has a completely different meaning compared to its use in regular expressions.) We assume that the alphabet is {a, b, c, d, e}.

*Solution.*

$$((a|b|c|d|e)*a(a|b|c|d|e)b(a|b|c|d|e)*)$$

$$((a|b|c|d|e)*ab(a|b|c|d|e)*cd(a|b|c|d|e)*) \qquad \square$$

**10.7.9.** Prove that for any regular expression there exists a finite automaton that recognizes the corresponding set of strings.

*Solution.* To prove this, we need the notion of a *nondeterministic finite automaton*. Consider a directed graph containing several points (vertices) and some arrows (edges) connecting those points. Assume that some of the edges are labeled by letters (from a given alphabet) and some edges remain unlabeled. Assume also that two vertices are selected; one is called the *initial* vertex I and the other is called the *final* vertex F. Such a labeled graph is called a nondeterministic finite automaton.

Let us consider all the paths from I to F. Going along a path, we read all the letters (on labeled edges). Therefore, each path from I to F determines a string. The automaton as a whole determines a set of strings, namely, the set of all strings that can be read along some path from I to F. We say that these strings are *accepted* by the automaton.

*Remark.* If we draw the states of a finite automaton as points and the transitions as labeled edges, it is clear that finite automata are special cases of nondeterministic finite automata. They are distinguished by the following requirements: (a) all edges are labeled except for the edges directed to the final vertex; (b) for each vertex and for each letter there is exactly one outgoing edge labeled by this letter.

We transform a regular expression into a finite automaton in two stages. First, we construct a nondeterministic finite automaton that corresponds to the same set. Then for any nondeterministic finite automaton we construct an equivalent deterministic finite automaton.

**10.7.10.** A regular expression is given. Construct a nondeterministic finite automaton that corresponds to the same set.

*Solution.* This automaton is constructed inductively, following the definition of a regular expression. If the regular expression is a letter or $\varepsilon$, the corresponding automaton has one edge. If the regular expression is $\Lambda$, the automaton has no edges at all. A union is implemented as follows:

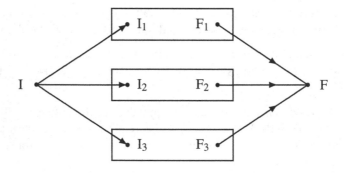

Here the picture for the union of three sets is drawn. The rectangles show the corresponding nondeterministic finite automata; their initial and final vertices are shown. New arrows (there are six of them) are unlabeled.

Concatenation corresponds to the following picture:

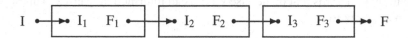

Finally, iteration corresponds to the picture

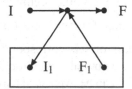

$\square$

**10.7.11.** A nondeterministic finite automaton $N$ is given. Construct an equivalent deterministic finite automaton (or a program with a finite number of states) that checks if an input string $x$ is accepted by $N$ (that is, if $x$ can be read on a path from I to F).

*Solution.* The states of the deterministic automaton are *sets* of vertices of the nondeterministic automaton. After a prefix $X$ of the input string is read, the state $s(X)$ of the deterministic automaton is the set of all vertices that are reachable from I along paths carrying the string $X$ on it. In other words, consider all paths starting from I. Each path determines a string that can be read along it. If the string is $X$, include the end of the path into $s(X)$.    $\square$

The two-stage construction of a finite automaton corresponding to a given regular expression, is finished.    $\square$

It turns out that regular expressions, deterministic finite automata, and nondeterministic finite automata define the same class of sets. To prove this, it remains to solve the following problem:

**10.7.12.** A nondeterministic finite automaton is given. Construct a regular expression that defines the same set.

*Solution.* Assume that the nondeterministic automaton has vertices $1, \ldots, k$, where 1 is its initial vertex and $k$ is its final vertex. By $D(i, j, s)$ we denote the set of all strings read along all the paths from $i$ to $j$ if only $1, 2, \ldots, s$ are allowed as intermediate path vertices. By definition, the automaton itself corresponds to the set $D(1, k, k)$.

We prove by induction over $s$ that all sets $D(i, j, s)$ for all $i$ and $j$ are regular. For $s = 0$, this is obvious (intermediate vertices are not permitted, therefore each set is a finite set whose elements are strings of length not exceeding 1).

Which strings are elements of $D(i, j, s + 1)$? Let us consider a path from $i$ to $j$ and mark all the steps when it enters the $(s + 1)$-th vertex. The marked steps split our path into several paths that do not use $s + 1$ as an intermediate vertex. This argument leads to the equation

$$D(i, j, s + 1) = D(i, j, s) \mid (D(i, s + 1, s)\ D(s + 1, s + 1, s)*\ D(s + 1, j, s))$$

(here the notation for regular expressions is used for sets). It remains to apply the induction assumption.                                                                                 □

**10.7.13.** Where have you seen a similar argument?

*Solution.* In the Floyd algorithm for the shortest path (see chapter 9, p. 130).     □

**10.7.14.** Prove that the class of sets corresponding to regular expressions remains the same if we agree to use not only set union but also complementation (and therefore set intersection, since it can be expressed using set union and complement).

*Solution.* For the deterministic finite automata the transition from a set to its complement is evident.                                                                           □

*Remark.* From a practical point of view, things are not so easy. The problem is that the transition from a nondeterministic automaton to a deterministic one may exponentially increase the number of states. There are many theoretical and practical questions concerning this problem. See the book of Aho, Sethi, and Ullman on compilers [2].

## 10.8 Suffix trees

Until now our programs first get a pattern that we are looking for and then the text to search in. In the following problems it is the other way around.

**10.8.1.** The program gets the input string $Y$ and can process it (no time or space limits yet). Then it gets the string $X$ of length $n$ and must report whether $X$ is a substring of $Y$. The number of operations in processing $X$ must be $O(n)$ (it must not exceed $cn$ where $c$ may depend on the alphabet). Construct such a program.

*Solution.* While processing $Y$ is unrestricted in time and space, it is not difficult. Specifically, one must "glue together" all the substrings of $Y$ into a tree grouping the common prefixes (as we did when matching several patterns at the same time). For example, $Y = ababc$ produces the following tree (vertices correspond to the substrings of $Y$, edges are labeled by letters added while traversing the edge).

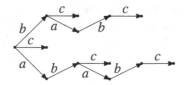

After such a tree is constructed, we read $X$ from left to right and try to follow the corresponding path in the tree starting at the root. String $X$ is a substring of $Y$ if we can do this while not leaving the tree. □

Note that this trick can be used for any set of strings $U$ (not just for the set of substrings of a given string). After constructing the tree, we can find out for any string $X$ whether it is in $U$ in time proportional to the length of $X$. (We need to keep some additional information indicating whether current vertex represents an element of $U$ or only a prefix of some element.)

**10.8.2.** Solve the previous problem with the additional restriction: the space used should be proportional to the length of $Y$.

*Solution.* The previous method does not work: the number of vertices of the tree is the number of substrings of $Y$ and can be proportional to $m^2$ for a string of length $m$.

However, we can "compress" our tree, leaving only the branching points. The edges are then labeled by substrings of $Y$ instead of characters.

Here is a compressed version of our example:

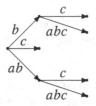

Let's assume (from now on) that the last character of $Y$ occurs only once in $Y$. (We can achieve this by adding an ad hoc terminator symbol at the end.) Then the leaves of a compressed tree are the suffixes of $Y$ and the internal vertices represent substrings of $Y$ that appear in $Y$ more than once followed by different characters. Each internal vertex (non-leaf) of a compressed tree has at least two sons. This implies that the number of internal vertices does not exceed the number of leaves. (In general, the number of leaves in any tree is greater than the number of internal vertices that have more then one son. When a new son is born, it either becomes a leaf instead of its father, or increases the number of vertices that have more than one son.) Since there are $m$ leaves there must be at most $2m$ vertices and the memory used is proportional to $m$, as long as we mark the edges in a clever way. Each mark is a substring of $Y$, so we represent it by two integers (the positions of its starting point and end point in $Y$). This doesn't make it harder to trace an arbitrary string $X$ in this tree character by character, just sometimes we are within the edges (and must remember the edge and the position). Reading a next character of $X$, we compare it with the corresponding character of the edge label (this can be done in $O(1)$ steps as the position of the character in $Y$ is known). □

The tree constructed in this way is called a *compressed suffix tree* for a string $Y$.

**10.8.3.** Prove that the compressed suffix tree can be constructed in time $O(m^2)$ and space $O(m)$ for a string of length $m$.

*Solution.* Let us construct the compressed suffix tree adding the suffixes one at a time. Adding one suffix is a task that is similar to tracing a substring: we read this suffix letter by letter and find the corresponding path in the tree. At some point the suffix leaves the tree (recall our assumption of the unique last symbol).

If the departure point is in the middle of an edge label, we have to cut the edge (and its label) into two pieces; a new branch point appears together with a new leaf that is a son of this point. If the departure point coincides with one of the existing vertices, this (internal) vertex gets a new son. Anyway, we need only $O(1)$ operations to restructure the tree (in particular, it is $O(1)$-easy to cut the label since it is represented by its start/end positions).

It is much more difficult to construct a compressed suffix tree in linear time (instead of quadratic time, as in the previous problem). We explain a McCreight algorithm that does this job. Let us start with some preparations.

First of all, let us describe in more detail the structure of the trees we are working with and some operations on these trees. We consider rooted trees whose edges are labeled by substrings of some fixed string $Y$. These trees satisfy the following conditions:

- each internal vertex has at least two sons;
- for any (internal) vertex all the outgoing edges have labels with different first letters.

Each vertex $v$ of such a tree corresponds to a string that can be read on the way from the root $r$ to $v$. We denote this string by $s(v)$. A label on the edge that goes to vertex $v$ is denoted by $l(v)$; the father of $v$ is denoted by $f(v)$. Then we can say that $s(r)$ is an empty string $\Lambda$ and

$$s(v) = s(f(v)) + l(v)$$

for every vertex $v \neq r$. (Here the "+" sign stands for the concatenation of strings.)

We consider not only vertices of the tree, but also *positions* in the tree; a position can be located in the tree vertex but can also be located "inside its edge" splitting the edge label into two nonempty parts. More formally, a position is a pair $(v, k)$ where $v$ is some non-root vertex of the tree and $k$ is an integer number in the range $[0, |l(v)|)$ that indicates how many steps back to the root we should do starting from $v$ before we reach this position. Here $|l(v)|$ is the length of the label $l(v)$; the value $k = |l(v)|$ would correspond to the preceding vertex in the tree and therefore is not allowed. In addition, we consider the position $(r, 0)$ that corresponds to the root of the tree. For each position $p = (v, k)$ we denote by $s(p)$ the corresponding string (that can be obtained from $s(v)$ by deleting $k$ last characters).

Let $p$ be an arbitrary position in the tree and let $w$ be a string. To *trace $w$ starting from $v$* means to find another position $q$ such that $s(q) = s(p) + w$. If such a position exists, it can be found in time proportional to the length of $w$. (Recall that a label is represented by the positions of its starting point and end point in $Y$.) If a position $q$ with required $s(q)$ does not exist in the tree, we need to "leave" the tree at some step. At this moment we enhance the tree making the remaining part of $w$ (which initially is outside of the tree) a label on the edge to new leaf. For that we need the remaining part of $w$ to be a substring of $Y$ (due to the representation of labels); it is guaranteed if the string $w$ itself is a substring of $Y$.

This process may create a new vertex in the tree or not, depending on whether the new edge starts from a vertex of the old tree or from a point on its edge. The number of operations is proportional to the length of the part of $w$ that is inside the old tree. (Note that the length of the second part of $w$ that remains untraced does not matter.)

It turns out that we may speed up the navigation in the tree *if we know in advance that it would be successful.*

**10.8.4.** For a given position $p$ and string $w$ there exists a position $q$ such that $s(q) = s(p) + w$; show that $q$ can be found in time proportional to the number of edges on the path from $p$ to $q$. (This number could be significantly smaller than the length of $w$ if edge labels are long.)

*Solution.* Indeed, the navigation operations are needed only in the vertices (we need to choose an outgoing edge with the correct first letter); in other cases the path is determined uniquely and we can jump to the end of the edge. □

So what have we achieved? We developed an efficient way (for a string $Y$) to store trees whose edges are labeled by substrings of $Y$ and perform the following operations:

- make a tree consisting of the root [$O(1)$];
- find the father of a non-root vertex [$O(1)$];
- find the label of a non-root vertex, i.e., the label of its incoming edge [$O(1)$];
- trace a string $w$ (knowing we will not leave the tree) starting from a position $p$; the operation returns a position $q$ such that $s(q) = s(p) + w$ [$O$(number of edges followed)];
- add a string $w$ that is a substring of $Y$, starting at position $p$; if a tree does not have a position $q$ such that $s(q) = s(p) + w$, such a position is created (it is a leaf) [$O$(the number of letters in $w$ that are not included in $l(q)$)];
- check whether a string $X$ appears in the tree, i.e., $X = s(q)$ for some position $q$ [$O(|X|)$].

The number of steps required for each operation is shown in square brackets.

This is still not enough to construct the compressed suffix tree for $Y$ in an efficient way. We need also to store the "suffix links" in the vertices of our tree (each vertex has at most one link) but let us first explain them.

Consider first the uncompressed suffix tree for a string $Y$. Each vertex of this tree (except the root) represents a nonempty substring of $Y$. If we delete the last character in this substring, we make one step toward the root of the tree. What if we delete the *first* character? Then we get a substring which may be in a more distant place of the tree.

The following picture shows the corresponding "jumps": each dashed arrow represents deleting the first character:

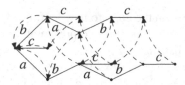

We call these arrows *suffix links* since they go from a string to its suffix (with the first character deleted). They are defined for all vertices in the suffix tree except the root.

More formally, let $w'$ denote $w$ with the first character deleted, so that $w'$ is defined for every nonempty string $w$. Then the suffix link goes from vertex $p$ to vertex $q$ if $s(q) = s(p)'$. (Recall that $s(u)$ is the string that corresponds to the vertex $u$.)

**10.8.5.** Let $u$ be a father of $v$ in the uncompressed suffix tree for some string $Y$. What can be said about the suffix links for $u$ and $v$?

*Answer.* They lead to a father-son pair of vertices $u'$ a father of $v'$, with the same character on the connecting edge.                                            □

Now we switch to the compressed suffix trees.

**10.8.6.** Prove that for a compressed suffix tree the suffix links still go from vertex to vertex (i.e., there is no suffix link that goes from a vertex to a position inside an edge).

*Solution.* As we assume that the last character of $Y$ is not recurrent, a suffix link that starts in a leaf ends in another leaf. A suffix link that starts in a branching vertex corresponds to a string $s$ which occurs in $Y$ several times with different characters after it: there exist two different letters $a$ and $b$ such that both $sa$ and $sb$ appear in $Y$. Deleting the first letter in $sa$ and $sb$, we get two substrings $s'a$ and $s'b$ of $Y$, so that $s'$ is also a branching point (and thus a vertex of the compressed suffix tree).          □

This is what our example looks like:

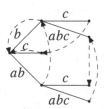

Now we are ready to present McCreight's algorithm. The compressed suffix tree is constructed step by step: we add suffixes in the order of decreasing lengths. Let $m$ be the length of $Y$ and let $Y_i$ denote the suffix beginning with the $i$th character of $Y$ (e.g., $Y_1 = Y$ and $Y_m$ consists of one letter). After $i$ steps, the tree stores $Y_1, \ldots, Y_i$.

**10.8.7.** Show that suffix links are well defined in this tree (lead to another vertex of it) for every vertex with two possible exceptions: the last added suffix $Y_i$ (the vertex that represents it) and its father.

*Solution.* The suffix link goes from the leaf $Y_j \neq Y_i$ to the leaf $Y_{j+1}$. Let $v$ be an internal vertex which is not father to $Y_i$ (the last leaf). The vertex $v$ is a branching point; let $Y_j$ and $Y_k$ be two leaves whose paths branch at $v$. We may assume without loss of generality that $j, k < i$ (if one of the paths goes to the last added leaf $Y_i$, it can be replaced by a path to another leaf: since $v$ is not the father of last added leaf, it is not the last branching point in the path). When we delete the first character from $Y_j$ and $Y_k$, we get $Y_{j+1}$ and $Y_{k+1}$. Corresponding paths are already in the tree since $j + 1, k + 1 \leqslant i$. The point where these two paths diverge is the endpoint of the suffix link from $v$.                                                                                 □

Suffix links need not be found for the leaves but are calculated (and stored) for all other vertices. After $i$ steps of the algorithm:

- the tree contains $Y_1, \ldots, Y_i$ (and all their prefixes);
- the address of the leaf which corresponds to the last added suffix $Y_i$ is stored in the variable *last*;
- all the internal vertices of the tree, except for the father of *last* (maybe), store a valid suffix link.

How can we maintain these properties while adding new suffixes? The naïve approach would be to add $Y_{i+1}$ character by character, starting from the tree root. (This led to having quadratic time previously.)

What optimizations are possible? There are two of them. First: we can move faster along the tree if we know that we are guaranteed to stay inside the tree. Second: we can use suffix links.

Both optimizations assume that the father of the leaf *last* (we denote this father by $u$ in the sequel) is not the root vertex. (If it is, we need to add the suffix $Y_{i+1}$ starting from the root.) Let *tail* be the label of the leaf *last* and let *head* $= s(u)$; in other terms, the string *head* corresponds to the vertex $u$. Then $Y_i = head + tail$.

Deleting the first letter, we get $Y_{i+1} = head' + tail$. Note that $head'$ is guaranteed to be inside the tree. Indeed, $u$ was a branching point, so not only the suffix $Y_i$ passes through $u$ but also some $Y_j$ for $j < i$. Then $Y_j$ has *head* as a prefix, and $Y_{j+1}$ starts with $head'$ and is already present in the tree.

So we may start by tracing $head'$ and find a position $v$ such that $s(v) = head'$, and then add *tail* starting from $v$.

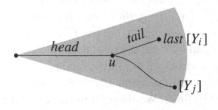

**Fig. 10.1.** First optimization: $head'$ is guaranteed to be in the tree.

This optimization does not use suffix links at all. They are used in the second optimization trick; this trick allows us (when applicable) to avoid tracing $Y_{i+1}$ from the root (even in its optimized version). Assume that the path to the leaf *last* (that corresponds to $Y_i$) goes through a vertex $v$ whose suffix link points to a vertex $w$ such that $s(w) = s(v)'$. Let $p$ be the string on the path from $v$ to *last*. Then

$$Y_i = s(v) + p, \quad Y_{i+1} = Y_i' = s(v)' + p = s(w) + p,$$

and to add $Y_{i+1}$ into the tree it is enough to add the string $p$ starting from vertex $w$.

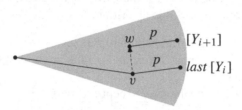

**Fig. 10.2.** Second optimization: we use the suffix link of a vertex $v$ that is on the path to *last*.

Both optimizations can be combined: we may trace the part of the string $p$ from $v$ to the father of the leaf *last* knowing in advance that we will stay in the tree.

Now we can describe the actions needed to add the next suffix $Y_{i+1}$ to the tree we construct, in the following way:

Let $u$ be the father of the leaf *last* that corresponds to the last suffix added to the tree.

Case 1: $u$ is the tree root. Then no optimization is possible and we add $Y_{i+1}$ starting from the root.

Case 2: $u$ is not the tree root but the father of $u$ is a tree root (i.e., the leaf *last* is two levels above the root). Then $Y_i = head + tail$ where *head* and *tail* are the labels of vertices $u$ and *last*. Then we use the first optimization trick and trace *head'* to some vertex $z$ (being sure that we do not get out of the tree); then we add *tail* starting from $z$.

Case 3: $u$ is not the tree root and its father $v$ is not the tree root either. Then $v$ has a valid suffix link to some vertex $w$ such that $s(w) = s(v)'$. Let *pretail* be a label of the vertex $u$ and let *tail* be the label of the leaf *last*. Then $Y_i = s(v) + pretail + tail$ and, therefore, $Y_{i+1} = Y_i' = s(w) + pretail + tail$. It remains to trace the string *pretail* starting from $w$ (being sure to remain in the tree) and then add *tail* starting from $z$.

It remains to understand how we can maintain the structure of suffix links (there is no problem with the address of the newly added leaf, we get it for free), and estimate the total time needed to perform all these operations.

Let us start with suffix links. We want them to be valid for all internal vertices (except for the father of the newly added leaf). So we need to take care of the father of the leaf *last* that corresponds to $Y_i$, because this leaf is no more freshly added. (Note that the only new vertex is a father of the leaf that was just added, so we do not need to care about it.) This is necessary in Case 2 and Case 3, but in these cases we have found the vertex $z$ that becomes an endpoint of the new suffix link. More precisely, $z$ could be not a vertex, but a position; but then it is transformed into a vertex (the father of the leaf that we add) during the operation. In the new tree $u$ is no longer a father of the newly added leaf and therefore (as we have seen) the suffix link for $u$ must point to a vertex. (In other words, if we are in Case 2 or Case 3 and position $z$ is inside some edge, then this edge is cut at $z$.)

This leads to the following procedure that adds the next suffix $Y_{i+1}$:

```
        { tree contains the suffixes Y₁, ..., Yᵢ;
            s(last) = Yᵢ;
            valid suffix links exist for all the internal
            vertices, except for the father of the leaf last }
        u := father of last;
        tail := label of last;
        { Yᵢ = s(u) + tail }
        if u = tree root then begin
          { Yᵢ₊₁ = tail' }
          add tail' starting from the root
              store the resulting leaf in last
        end else begin
          v := father of u;
          pretail := label of u;
          { Yᵢ = s(v) + pretail + tail }
          if v = tree root then begin
            { Yᵢ₊₁ = pretail' + tail }
            trace pretail' from the root to z
          end else begin
            w := suffix link for v;
            { s(w) = s(v)', Yᵢ₊₁ = s(w) + pretail + tail }
            trace pretail from w to z
          end;
          { it remains to add tail starting from z and the link from u to z }
          if position z is a vertex then begin
            add a link from u to z;
            add tail starting from z,
                put the resulting leaf into last;
          end else begin
            add tail starting from z,
                put the resulting leaf into last;
            add a link from u to the father of last;
          end
        end;
      end;
```

It remains to estimate the time needed for adding all suffixes $Y_1, \ldots, Y_m$ one after another. Each next suffix requires $O(1)$ time if we do not count the time needed for "tracing" and "adding" things. We need to prove that the total number of operations is $O(m)$, so it is enough to show separately that the total time needed for tracing and adding operations is $O(m)$. (Note that this does not prevent some tracing or adding operations to be long if others are short.)

**Tracing operations.** The time needed to perform a tracing operation is proportional to the number of the edges involved (we denote this number by $k$). Note that the height of the last added leaf increases by $k - O(1)$ (compared to the leaf that was the newly added one before the operation). Indeed, note that in Case 3 the height of vertex $s(w)$ could be less than the height of $s(v)$ at most by one, since suffix links for all vertices on the path to $v$ (except for the root that has no suffix link) point to vertices on the path to $w$.

Since leaves always have height at most $m$, the total time needed for all tracing operations is $O(m)$.

**Addition operations.** We use a similar argument but replace the height of a newly added leaf by the length of its label. Adding the string *tail* (or *tail'*) requires time proportional to the number of characters we read, but each of them (except one, may be) decreases the length of the label at least by 1: the new label consists of unread characters (except for the first one). Therefore all addition operations require (in total) $O(m)$ time.

Therefore we have shown that the described algorithm constructs the compressed suffix tree for a string $Y$ of length $m$ in time $O(m)$. After that we need only $O(n)$ time to find out whether a given string $X$ of length $n$ is a substring of $Y$.

**10.8.8.** How should we modify this algorithm not only to get a "yes/no" answer but also to find the position where $X$ appears in $Y$ (if the answer is "yes")? If there are several occurrences of $X$ in $Y$, one position is enough. The time should remain $O(|Y|)$ for preprocessing $Y$ and $O(|X|)$ for processing $X$ after that.

*Solution.* Every vertex of a compressed suffix tree corresponds to some substring of $Y$. When this vertex was added, the information about the position where this substring occurs was available, and the endpoint of this substring can be recorded in the vertex.                                                                              □

**10.8.9.** How should we modify the algorithm in such a way that we can find the *first* (= leftmost) occurrence of each substring?

[*Hint.* When a new vertex appears inside the edge, we should use its first occurrence (the corresponding information is available at the end of the edge), not the just discovered one.]                                                                              □

**10.8.10.** How should we modify this algorithm in such a way that we can find the *last* (= rightmost) occurrence of each substring?

[*Hint.* We cannot afford the time needed to update the information along the path on each step; a more effective solution is to construct the entire tree and then traverse it again computing for every vertex the last occurrence of the corresponding substring.]                                                                              □

**10.8.11.** How can we use a compressed suffix tree to find, for a given string $Y$, the longest substring that occurs more than once? The time should be $O(|Y|)$.

[*Hint.* Such a substring is an internal vertex of a compressed suffix tree, so it is enough to choose its vertex that corresponds to the longest substring. For that it is enough to traverse the tree (we can compute the length on the fly by adding and subtracting the lengths of labels).] □

Practically it may be simpler (and even faster in some cases) to use another algorithm to find the longest substring that occurs more than once. This trick is called *suffix array*. Let us consider the number $i$ as a "name" for the suffix of the string $Y_i$ that starts with $i$th letter. We introduce ordering on names that corresponds to the lexicographic order of the suffixes: $i$ is "less" than $j$ if $Y_i$ precedes $Y_j$ in the lexicographic order. Then we sort all the names according to this order and get some permutation of $1, 2, \ldots, m$ where $m$ is the length of $Y$. If some string $X$ appears in $Y$ twice, then $X$ is a prefix of two suffixes of $Y$. We may assume without loss of generality that these suffixes are neighbors in the lexicographic order (all the intermediate elements also have prefix $X$). Therefore it is enough to find the maximal common prefix between neighbor elements of the sorted array.

This approach requires less space (only one auxiliary array of $m$ integers) but may require more time: first, sorting itself requires $m \log m$ comparisons; second, each comparison may require a lot of operations (if the common part is long). But if there are few coincidences, this approach is quite practical.

**10.8.12.** Apply one of these algorithms to your favorite book and explain the result.

[*Hint.* The long parts that appear more than once could be intentional (see, e.g., "This is the house that Jack built...."). In modern books they could mean that author abuses the cut and paste operations while using text editor.] □

# 11

# Games analysis

In this chapter we consider a basic notion of game theory: a class of games called finite perfect information games. First (section 11.1) we consider some examples that illustrate the notion of a winning strategy. Then in section 11.2 we prove the Zermelo theorem and define the notion of game cost. This leads to an algorithmic question: how can we compute the game cost? In section 11.3 we show an algorithm based on the full traversal of the game tree, and in section 11.4 we study an optimization technique that allows us to compute (exactly) the game cost avoiding some parts of the game tree. Finally, in section 11.5 we apply dynamic programming to game analysis.

## 11.1 Game examples

**11.1.1.** There are 20 matches on the table. Two players make alternating moves; each could take any number of matches between 1 and 4. The player who cannot make a move (as there are no more matches) loses the game. Who wins the game if both players play optimally?

*Solution.* The second player wins by complementing the opponent's move so that both take 5 matches in total. (If the first player takes 1 match, the second should take 4, etc.) After four rounds there are no matches (each round deletes 5 matches) and the first player loses the game.                                                                          □

**11.1.2.** What if the same game is started with 23 matches?

*Solution.* The first player wins: if she takes 3 matches at her first move, then she plays the role of a second player in the game with 20 matches and can use the strategy described above.                                                                                     □

For the same reasons the second player wins if game starts with $N$ matches where $N$ is a multiple of 5; otherwise the first player wins.

**11.1.3.** Let us change the rules of the game: the player who takes the last match loses. Who wins if both players do their best? (The answer depends on $N$.)      □

A. Shen, *Algorithms and Programming*, Springer Undergraduate Texts in Mathematics and Technology, DOI 10.1007/978-1-4419-1748-5_11,
© Springer Science+Business Media, LLC 2010

**11.1.4.** Consider a game with similar rules: each player may take 1, 2, or 4 matches and loses if are no matches. Who wins if a game is started with 20 matches (and both players play optimally)?

*Solution.* Here it is more difficult to find the optimal strategy. Let us start with small examples. We show allowed moves in a diagram (Fig. 11.1; moves are represented by arcs): A player that finds 0 matches before her move, loses (according to

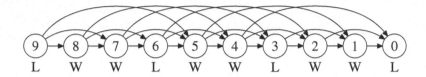

**Fig. 11.1.** Game with matches.

the rules). So we write "L" under the 0-circle. Therefore, the player that has 1, 2 or 4 matches before her move, wins: she can take all the matches and put her opponent in the losing position (with 0 matches). So we write "W" under positions with 1, 2 and 4 matches. Now it becomes clear that 3 is a losing position: one can take 1 or 2 matches but this leaves the opponent in a winning position (both 1 and 2 are winning positions). So we write "L" under 3. This implies that 4, 5 and 7 are winning positions since the player can take 1, 2 and 4 matches (respectively) and let the opponent play with 3 matches. And then we see that 6 is a losing position since all arcs go to winning positions 5, 4, 2; then 7 and 8 are winning positions, 9 is a losing position and so on with period 3. All the positions where the number of matches is a multiple of 3 are losing ones. All the other positions, including 20, are winning positions.  □

**11.1.5.** What is the winning strategy for the first player?

*Solution.* She should put her opponent in the losing position, i.e., maintain the following invariant relation: after her move the number of matches is a multiple of 3. She should start the game by taking 2 matches (18 remain, and 18 is a multiple of 3).  □

**11.1.6.** The game starts with two groups of $m$ and $n$ matches. The players take turns taking any number of matches *from one group*. (E.g., the player may remove the entire group.) A player that cannot make the move (no matches remain) loses the game. Who wins if both players play optimally?

*Answer.* If $m = n$, the second player wins; if $m \neq n$, the first player wins.  □

**11.1.7.** A rook is placed on the chess board; two players make alternating moves. Each player moves the rook left or down. A player that cannot make a move loses the game (this happen when the rook is in a1 position). Who wins if both players do their best?

[*Hint.* What is the connection between this game and the previous one?]  □

**11.1.8.** The game starts with $k$ groups of $n_1, \ldots, n_k$ matches; two players take turns removing matches; at each move, a player can remove any number of matches from one group (and may remove the group altogether). The player who cannot make a move (no matches left) loses the game. Who wins if both players do their best? (This game is sometimes called *Nim*.)

*Solution.* Let us write the numbers $n_1, \ldots, n_k$ in binary, as bit strings, with last digits aligned (as if we were adding them). We claim that if each column has an even number of ones, then the second player wins; if there is a column that has an odd number of ones, the first player wins. Indeed, if each column has an even number of ones and the number in some row decreases, some bit in this number is changed and therefore some column gets an odd number of ones. (This corresponds to the fact that every move from a losing position ends in a winning position.) Assume that some column has an odd number of ones. Let us take the leftmost column with this property and consider one of the numbers that has 1 in this column. Replace this 1 by 0: this makes the number smaller even if you change arbitrarily the bits on the right of this 1. This allows us to make all columns even: the column we started with becomes even automatically, and for all columns on the right of it we choose the proper bit in the row we change. (See the rules for winning and losing positions in the next section.) □

**11.1.9.** There are $n$ glasses in a row; each contains a coin. Players take their turns by taking out one or two coins with the following restrictions: if you take two coins, they should be in the neighbor glasses. Player who cannot make the move (no coins left) loses the game. Who wins in this game if both players play optimally?

*Solution.* The first player can win by taking one or two coins in the middle thus splitting the glasses in two symmetrical groups. Then she should repeat the moves of the opponent in the other group. □

## 11.2 Game cost

While analyzing the games in the previous section and classifying the positions into *winning* and *losing* ones (for the player that finds herself in this position), we used the following (evident) rules:

1. If there exists an arc that goes from some position $p$ to a losing position, then $p$ is a winning position.

2. If all the moves that are possible in position $p$ lead to winning positions, then $p$ is a losing position.

**11.2.1.** Consider a game that has finitely many positions and no loops (we cannot return to a previously visited position). Assume that for every terminal position (where no moves are possible) the winner is fixed (it is known whether a player that ends in this position is a winner or a loser). Prove that rules 1 and 2 uniquely classify all positions into winning and losing ones.

*Solution.* Let us apply these rules wherever possible. It is clear that no position can be declared as winning and losing at the same time (rule 1 can be applied only if rule 2 cannot). We need to prove that there are no "uncertain" positions (assuming that rules cannot be applied anymore). Note that for every uncertain position there is at least one arc that leads to another uncertain position. (If all arcs go to positions that are declared as winning or losing, one of the rules 1 and 2 can be applied.) Following these arcs, we find an uncertain path that at some point should become a loop. □

**11.2.2.** A similar statement is true for games that allow draws. Give a precise statement and prove it. □

Games with draws are a special case of a general notion: a two-person *zero-sum game*. To specify a zero-sum game, we:

1. fix a finite set whose elements are called *positions*;
2. for each position we say whether it is *terminal* (the game is over) or not;
3. for each terminal position we specify some number; it is interpreted as the amount of money that (say) the first player pays to the second one;
4. for each non-terminal position we specify which player makes a move in this position and which moves are allowed (what are the possible positions at the next step);
5. specify a starting position.

The position graph should have no loops (the game cannot return to the same position after several moves).

The position are vertices of some directed graph, and moves are its arcs (edges). One can draw the graph of the game on a specially designed game board; then players move a token from vertex to vertex according to the rules of the game. At each non-terminal vertex it is written who should make the next move; at each terminal vertex it is written how much one of the players should pay to the other one. One of the vertices is labeled as an initial vertex.

Let us call the players Max and Min; after the game is finished, Min pays the amount (indicated by the game rules) to Max. (This explains the name: Max wants to maximize the amount, and Min wants to minimize it.) Both Max and Min can start the game (depending on what is said in the initial vertex). Note that we do not assume that players alternate: the same player can make several moves in a row if the vertices labels dictate this.

An important remark: the graphs we have drawn (say, for the matches game) are not the game graphs since it is not indicated in a vertex whether Max or Min makes a move there. To get a correct game graph, we should split each vertex in two: one for Max and the other one for Min. The arcs go across these two copies (since the players alternate). The winner and loser in a terminal position can be represented as numbers: $+1$ means that Max wins (and gets a unit payment from Min); $-1$ means that Min wins.

Let us define a notion of a *strategy* (for some player) in an arbitrary zero-sum game. A strategy for Max (or Min) prescribes a move for every position where Max (Min) should make a move. Formally a strategy for Max (Min) is a function $s$ defined for every position where Max (Min) should make a move, and $s(p)$ is one of the possible moves, i.e., there is an arc from $p$ to $s(p)$.

Such a strategy is often called a *positional strategy* since the move depends only on the current position (but not on the history of the game). We do not need other strategies here, so we usually omit the word "positional".

If the strategies for Max and Min are fixed, the game and its outcome are determined. But the player should choose a strategy not knowing in advance the strategy of the opponent (only the moves that are already done by the opponent are visible). It turns out that it is still possible to find an optimal strategy:

**11.2.3.** Prove that for every game $G$ there exists a number $c$ and strategies $M$ and $m$ for Max and Min such that:
(1) Using strategy $M$, Max wins at least $c$ (whatever Min does);
(2) Using strategy $m$, Min loses at most $c$ (whatever Max does).

This $c$ is called *game cost*. Note that game cost is unique: the conditions (1) and (2) guarantee that Max has no strategy that guarantees winning more than $c$ (since this would not work against strategy $m$) and Min has no strategy that guarantees losing less than $c$ (this would not work against strategy $M$).

If the game has only two outcomes $+1$ (Max wins) and $-1$ (Min wins), the statement of this problem says that one and only one of the players has a winning strategy. This statement is often called the *Zermelo theorem*. If we add a third outcome, a draw, the statement says that either one of the players has a winning strategy ($c = 1$ means that Max has it, and $c = -1$ means that Min has it) or both have a strategy that guarantees at least a draw ($c = 0$).

*Solution.* Let $p$ be a position in $G$. Consider the game $G_p$ that has the same rules as $G$, but the starting position is now $p$. If $p$ is a terminal position, then the game is trivial: Min pays Max the indicated sum, and this sum is the cost of $G_p$. This is a starting point for a recursion that determines the cost of $G_p$ for all vertices $p$.

More precisely, consider the following recursive definition of a function $c$ on the game vertices:

- if $p$ is a terminal vertex, then $c(p)$ is the amount that Max pays Min when the game ends in $p$;
- if Max makes a move in $p$, then $c(p) = \max\{c(p')\}$ where the maximum is taken over all vertices $p'$ to which Max can move from $p$ according to game rules;
- if Min makes a move in $p$, then $c(p) = \min\{c(p')\}$ where the minimum is taken over all vertices $p'$ to which Min can move from $p$ according to game rules.

**Lemma.** There exists a unique function $c$ defined on all game positions that satisfies these three requirements.

**Proof** of the lemma uses induction over a parameter called *rank* and defined as a maximal number of moves that can be done starting from a given vertex. There are no loops, therefore the rank of each vertex does not exceed the number of vertices. Note also that each move decreases the rank by at least 1. Let us prove that for every $k \geqslant 0$ there exists a unique function defined on the vertices of rank at most $k$ and satisfying the requirements. For $k = 0$ it is evident. The induction step: assume that the function is defined on vertices of ranks $0, 1, 2, \ldots, k-1$. Then the inductive rule determines the values of this function on vertices of rank $k$ (since $c(p')$ are already defined). The lemma is proven.

Now we need to prove that $c(p)$ (defined inductively) is indeed the cost of the game $G_p$. Consider the following (positional) strategy for Max: being in vertex $p$, Max selects $p'$ for which $c(p')$ is maximal among all the possibilities (and is therefore equal to $c(p)$). Assume that Max follows this strategy. Then (whatever Min does) the value $c(q)$ at the current vertex $q$ never decreases during the game (this was the rule for Max's strategy; during Min's move this value cannot decrease by definition of the function $c$). Therefore the payment in the end of the game is guaranteed to be $c(p)$ or more. Similarly, if Min chooses the vertex $p'$ where $c(p')$ is minimal (=equal to $c(p)$), then the value of $c(q)$ never increases during the game and therefore Min has to pay at most $c(p)$ at the end.

So we have proven the generalized Zermelo theorem.                                           □

**11.2.4.** Consider the following game: two players (in turn) put their signs (crosses and naughts) in empty cells of a large square board (one sign at a time). To win, player must have five signs in a row, in a column or diagonally (and this is the end of a game). If all the cells are occupied and this does not happen, the game is a draw. Prove that the first player (that puts crosses) has strategy that guarantees at least a draw (prevents from losing).

*Solution.* Assume that such a strategy does not exist. Then Zermelo's theorem guarantees that the second player has a winning strategy. But the first player could use essentially the same strategy starting from the second move. Indeed, imagine that the first cross is put with a pencil (in an arbitrary cell), and the next crosses are put in ink (according to the winning strategy for the second player, as if the pencil cross does not exist).

Could the pencil cross create a problem for the first player? Yes, if the strategy tells the first player to put a sign in the cell already occupied by a pencil cross. But in this case the first player can overwrite the pencil cross with an ink one, and place a pencil cross in any free cell. (If there are no free cells, the position represents the winning position for naughts in the symmetric game, so the first player wins.)

Also the game may end prematurely if a pencil cross completes the 5-crosses sequence, but this is also an advantage for the first player.

Therefore we had shown that if the second player has the winning strategy, then the first player has it too — a contradiction (imagine two winning strategies playing against each other). So the second player does not have it, and (according to Zermelo's theorem) the first player has a strategy that guarantees at least a draw.   □

**11.2.5.** Prove that the cost of a game is equal to the payment made in one of the terminal vertices.                                                                                     □

**11.2.6.** Prove that Zermelo's theorem is a consequence of its special case where only two outcomes are possible (one of the players wins).

[*Hint.* Possible payments divide the real line into intervals. For every interval we define a "reduced game": a threshold in this interval is chosen; the payment below/above the threshold is considered as losing/winning the game.] ☐

**11.2.7.** Let $G$ be some game. Choose one of the terminal vertices and let us change the payment in this vertex replacing it by some number $c$. The resulting game is denoted by $G[c]$. Its cost is a function of $c$. What kind of function may appear in such a way?

*Answer.* The cost of $G[c]$ is the closest (to $c$) point of some interval $[a, b]$. This interval depends of the game $G$. (The endpoints of this interval could be infinite.) ☐

Here is one more example of a game where Zermelo's theorem implies the existence of a winning strategy for the first player (but in a non-constructive way: no explicit strategy is provided). This game, called Bridge-it, was invented by David Gale and appeared in M. Gardner's *Scientific American* column in 1958.

Consider a grid made of dotted lines that form a rectangle of height $n$ and width $n+1$ (Figure 11.2); neighbor vertices of this grid are connected by (dotted) segments

**Fig. 11.2.** Bridge-it game.

of unit length. Players take their turns: at each move the first player may blacken any dotted segment of unit length (drawing a solid line over it). The ultimate goal of the first player is to connect left and right side by a (solid) path. The goal of the second player is to prevent this by erasing the dotted segments: at each step the second player could erase one of them (that is not blackened yet) preventing the appearance of a solid line in this place. The game is over when all the segments are either blackened or deleted; the first player wins if the left and right sides of the rectangle are connected.

**11.2.8.** Use Zermelo's theorem to show that the first player has a winning strategy in this game.

[*Hint.* One may present a game in a more symmetric way by adding an orthogonal mesh for the second player (Figure 11.3): now the second player wants to connect the top and bottom of the new mesh, and the lines of the first and the second players cannot intersect (thus each segment of the new mesh prevents one of the segments of the old mesh from appearing). After the game is over (for every point there is a vertical or a horizontal segment going through it), one of the players wins (and two players cannot win in the same time): either left and right sides are connected, or top and bottom sides are connected. (A rigorous proof of this statement is not that easy though our topological intuition says that it should be true.)]                                          □

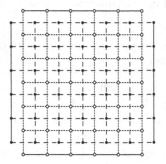

**Fig. 11.3.** A symmetric version of Bridge-it.

M. Gardner tells a story about this game: Claude Shannon (the inventor of information theory) suggested an interesting "physical" strategy that works for every grid. Replace all the dotted lines in the first player's grid by resistors (all having the same resistance); the right and left sides of the rectangle are conductors that have zero resistance. Darkening the line means short-cutting this resistor (reducing its resistance to zero); removing the dotted line means that the resistor is cut off (the resistance is made infinite). The Shannon's strategy for the first player is to apply some voltage between the left and right sides of the grid, find the most stressed resistor (for which the current and the voltage difference are maximal) and make the next move there. (If there are several resistors with the same current, one of them should be used.)

M. Gardner does not say whether this is indeed a winning strategy. Instead, he gives a winning strategy (attributed to Oliver Gross, see chapter 18 of his book *New Mathematical Diversions from "Scientific American"*, NY, 1966). To explain this strategy, let us recall that the goal of the first player is to prevent the second one from connecting the top and bottom side of the second grid. (As we have mentioned, this goal is equivalent to the original one.) The first move of the first player is shown in Figure 11.4; it destroys one of the edges of the second player. We split the remaining edges into pairs of neighbor edges as shown at the same drawing. The first player prevents the second one from creating both edges in any pair: if the second player created one of the edges, the first player responds by destroying the second edge in a pair (by creating an orthogonal edge). The next problem shows that this is indeed a winning strategy (that can be used by the first player to prevent the second one from making a top-bottom path which means that the first one creates a left-right path).

**11.2.9.** Prove that any path along the dotted grid (Figure 11.4) should include both edges of some pair.

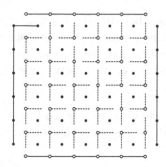

**Fig. 11.4.** A winning strategy for Bridge-it.

*Solution.* To make the picture easier to follow, let us use only dotted lines and corresponding vertices (Figure 11.5). Draw a gray area as shown in the picture; now edges are classified into white and gray edges. Assume there exists a path from bottom to top that does not cover any pair of edges. We may assume without loss of generality that this path never visits the same vertex twice (by deleting loops). Each step

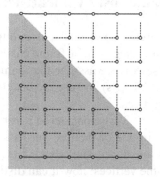

**Fig. 11.5.** The winning strategy for Bridge-it: the analysis.

in this path could belong to one of 8 classes: four directions (north=up, south=down, left=west, right=east) are combined with two colors (white/grey). The path starts with north-grey step and ends with north-white step.

Let us show that this is impossible if we do not violate the restrictions (do not use both edges of the same pair and do not visit one vertex twice). What can follow the north-grey step? There are three possibilities: one more north-grey step, west-grey step or east-white step. After the west-grey step only north-grey step or other west-grey step are possible. Looking at all the possibilities, we conclude that the path that starts with a north-grey step, never leaves the set

north-grey, west-grey, east-white, south-white

Therefore the north-white step (that should be the last step in the path connecting bottom and top) can not appear, and we get a contradiction. □

**11.2.10.** Two players take their turns by marking the edges of an infinite square grid (cell paper) by red and blue pencil. (The first player at each move can select any non-colored edge and make it red; the second player can do the same with blue color.) Prove that the second player may prevent the first one from creating a red loop.

[*Hint.* The second player may prevent the red loop to change a west direction to north direction by splitting all the edges in pairs and preventing the first player to use both edges in a pair.]                                                                □

The following problem (a classical result in percolation theory) uses the notion of probability.

**11.2.11.** Let us return to the Bridge-it field (Figure 11.2) and assume that each of the dotted segments is made solid with probability 1/2, and different segments are independent. Prove that a solid path between the left and right sides exists with probability 1/2.                                                                □


## 11.3 Computing the game cost by backtracking

The proof of Zermelo's theorem shows that it is enough to know the cost for all vertices (and then the optimal strategy can be found easily). How can we compute the costs of all vertices? In this section we assume that game graph is a tree (moves are edges that go from a vertex to its sons) and use the backtracking program (chapter 3) to compute the costs of all vertices.

Let us recall that in chapter 3 we considered a robot that walks over the tree. At any moment it is located in one of the tree vertices. The robot can perform the instructions up_left, right and down. The robot starts in the root. We have assumed that the robot can "process" the vertices; now it can distinguish between the vertices of different types (a type call to Robot returns max, min, or terminal) and can find the cost of a current vertex if it is a terminal one.

**11.3.1.** Write a program that uses the robot to collect the information and computes the cost of the game.

*Solution.* Let us write a recursive procedure that, being started at some vertex, traverses the subtree rooted at that vertex, brings the robot to the initial position and computes the cost of this vertex (where the procedure was called).

```
procedure find_cost (var c: integer)
 | var x: integer;
begin
 | if type = terminal then begin
 | | c:= cost;
 | end else if type = max then begin
 | | up_left;
```

```
find_cost (c);
{c is the maximum of the costs of the current
               vertex and all its left brothers}
while is_right do begin
   right;
   find_cost (x);
   c := max (c,x);
end;
{c is the cost of the father of the current vertex}
down;
end else begin {type = min}
   ...similar part where max is replaced by min
end;
end;
```

We use the fact that every vertex of type max or min has at least one son (the vertices that have no sons should be terminal vertices and have type terminal).  □

**11.3.2.** Write a non-recursive program that does the same (computes the cost of a game represented by a tree).

*Solution.* As usual, recursion can be replaced by a stack. Each element in the stack keeps information about some ancestor of the current vertex (the stack top keeps information about the father of the current vertex; the next element of the stack keeps the information about the grandfather, etc.) When the robot is in the root, the stack is empty. When the robot moves up in the tree, the stack becomes longer; when it moves down, the stack becomes shorter.

Each element of the stack is a pair. The first element of the pair is the type of the corresponding vertex $v$ (which is min or max; it cannot be terminal, since leafs are not ancestors and so do not appear in the stack). The second element of the stack is the maximal (resp. minimal) value among the costs of all the sons of $v$ that are on the left of the current son (on the path to the robot position).

Recall the two statements LP and LAP that played an important role in chapter 3: LP means that all the leaves on the left of the current vertex are processed; LAP means that all the leaves on the left and above the current vertex are processed (this happens when we go through the vertex for the second time).

In our program we not only use the stack but also some variable c. This variable is used only if LAP condition is true; in this case it contains the cost of the current vertex. Let us show how we can maintain this invariant relation while moving the robot along the tree, by extending the statements (1)–(4) made on p. 53:

- {LP, not is_up} process {LAP}:
  let c be the cost of the current leaf;

- {LP, is_up} up_left {LP}:
  before going up, we push into the stack the type of the current vertex (max or

min) and the value $-\infty$ or $+\infty$ respectively (the maximal value in an empty set is $-\infty$ and the minimal value is $+\infty$);

- {is_right, LAP} right {LP}:
  we update the value (second component) of the top element in the stack by performing max or min operation with the old value and c; the type of the operation is determined by the type of the top element in the stack (its first component);

- {not is_right, is_down, LAP} down {LAP}:
  the new value of c is max/min (the choice of the operation depends on the type of the stack top) of its old value and the value of the stack top; we perform the pop operation that makes the stack one element shorter.

It is easy to see that these actions maintain the invariant relation we have described, and when the tree traversal program terminates, the stack is empty and the cost of the entire game is contained in c.

(A pedantic reader would say that we make the invariant relation a part of LP and LAP conditions: both LP and LAP describe the contents of the stack and LAP also describes the value of c. On the other hand, "processing" leaves is not really needed, so we in fact replace LP and LAP by conditions described above.) □

**11.3.3.** How should we modify the program to compute the cost of all vertices in a game tree?

[*Hint.* When LAP is true, we know the cost of a current vertex, and this happens for every vertex.] □


## 11.4 Alpha-beta pruning

In the last section we have seen how one can compute the cost of a game by backtracking (visiting all vertices in a tree). However, sometimes it is not necessary to visit all vertices. Imagine that the game has two outcomes (we can win or lose) and we have found (after looking at some part of the tree) that some initial move is a winning one. Then we may ignore all other moves (not analyzed yet). More generally, if we have found a move that provides maximal possible gain (according to the game rules), we do not need to continue the analysis.

There are other cases when an optimization is possible. Consider the game tree shown in Figure 11.6. We assume that $a \geqslant b$ and we make our analysis from left to right. Then after analyzing $a$-vertex we know that the cost of the root vertex is at least $a$. Then we consider the min-vertex and after looking at its first (leftmost) son we know that the min-vertex cost is at most $b$. If $b \leqslant a$ we can be sure that the min-vertex does not change the cost of the root. So the grey part of the tree can be omitted. (If a chess master plans some move and find a very efficient answer to this move, she does not need to consider other answers.)

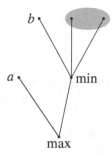

**Fig. 11.6.** A short-cut is possible if $a \geqslant b$.

How can we describe this kind of optimization in general terms? While starting to analyze some vertex, we know some interval $[m, M]$ that is interesting for us: either we know that the cost is guaranteed to be in this interval (e.g., this happens when all payments in the game are in $[m, M]$), or we know that the cost values outside $[m, M]$ do not really change anything (as in the second example where all costs less than $a$ are equivalent to $a$). This optimization is often called $\alpha$-$\beta$-*pruning* (since the bounds were once called $\alpha$ and $\beta$).

To make it more formal, let us introduce some notation. Let $x$ be a number and let $[a, b]$ be some interval. Then $x_{[a,b]}$ denotes the point in the interval that is the closest to $x$, i.e.,

$$x_{[a,b]} = \begin{cases} a, & \text{if } x \leqslant a; \\ x, & \text{if } a \leqslant x \leqslant b; \\ b, & \text{if } b \leqslant x. \end{cases}$$

We say that $x_{[a,b]}$ is "$x$ reduced to $[a, b]$". Now we can say that after analyzing the vertex $a$ (Figure 11.6) we are interested in the cost of min-vertex reduced to $[a, +\infty]$ (all values below $a$ are equivalent to $a$), and after looking at vertex $b$ this reduced cost is already known (equal to $a$). And for a game with two outcomes $-1$ and $+1$ the cost does not change being reduced to $[-1, +1]$, and after we find a winning move, we know that the cost is $+1$.

Using this observation, we write an optimized recursive algorithm that gets an interval $[a, b]$ and computes a cost of the current vertex reduced to $[a, b]$.

```
procedure find_reduced_cost (a,b: integer; var c: integer)
  var x: integer;
begin
  if type = final then begin
    c:= the cost of a current vertex reduced to [a,b]
  end else if type = max then begin
    up_left;
    find_reduced_cost (a,b,c);
    {c = the maximal value among the cost of the current
            vertex and its left brothers reduced to [a,b]}
    while is_right and (c<b) do begin
```

```
        |  | right;
        |  | find_reduced_cost (c,b,x);
        |  | c := x;
        | end;
        | {c=the cost of the father of the current vertex
        |                               reduced to [a,b]}
        | down;
      end else begin {type = min}
        | ...the symmetric part
      end;
  end;
```

The natural question now is whether this kind of optimization is really important. What part of the tree do we need to visit if we use this technique? To get some idea about that, let us consider a very simple example. Assume that the game has fixed length, the players alternate and every player makes $n$ moves. In every (non-terminal) position two moves are possible and the leaves have costs 0 and 1. Then the tree is a complete binary tree where min- and max-levels alternate, the leaves carry zeros and ones and we need to compute the value in the root. (If we identify 1 with `true` and 0 with `false`, the max- and min-operations become OR- and AND-operations, so trees of this type are sometimes called AND-OR-trees.).

How many leaves of an AND-OR-tree do we need to visit in order to compute the value in the root? Recall that the total number of leaves is $2^{2n}$ for the tree with $2n$ levels (each player makes $n$ moves), so we should compare the result with $2^{2n}$.

**11.4.1.** Prove that for any values in the leaves our optimized algorithm needs to visit at least $2^n$ leaves.

*Solution.* At level 2 the tree has four vertices. The algorithm should determine the cost of at least two of them. Indeed, let us assume that the root is min-vertex. If its cost is 0, then one of its sons has cost 0, and to establish this fact the algorithm needs to ensure that both sons of this son have cost 0. And if the cost of the root vertex is 1, then each son should carry 1. To establish this the algorithm should visit at least one son of each son.

For the same reason each value of level 2 needs at least two values of level 4, etc. So we need in total at least $2^n$ vertices of level $2n$.                                      □

In the preceding problem we considered a specific algorithm to compute the cost of the root. But one can prove a more general statement: any set of leaf values that uniquely determines the root value, consists of at least $2^n$ values.

**11.4.2.** Give a rigorous proof of this claim.

[*Hint.* In fact our proof gives just that, we need only to understand this.]        □

This was the lower bound for the best case. The next problem gives the lower bound for the worst case showing that our algorithm does not improve anything in the worst case.

**11.4.3.** Assume that somebody asks us what are the values in the leaves of an AND-OR-tree (in order which we do not control), and we can give whatever answers we want. Prove that we can choose our answers in such a way that the cost of the root remains unknown up to the last moment (i.e., could be 0 or 1 depending on the remaining leaves while at least one leaf remains).

This problem shows that any algorithm that (correctly) finds the root cost could be forced to check all the $2^{2n}$ leaves, so no optimization is possible in the worst case.

*Solution.* Let us use induction over the tree height. Assume that the root is an AND-vertex. By the induction assumption, we can leave the cost of a root's son not determined by the disclosed values until the last leaf above it is questioned. And when this happens (the last leaf in one of two subtrees is asked), we choose the leaf value to make this subtree true. So the root value is the value of the other subtree and remains unknown until the last moment. □

The more interesting question is the *average* number of leaves to be asked when computing the root cost. Imagine that find_reduced_cost is applied to some fixed AND-OR-tree (with fixed values of all leaves), but the algorithm is randomized in such a way that at any vertex the ordering of two its sons (which one should be visited first) is chosen by a coin tossing. Then the total number of visited leaves becomes a random variable (a function of the results of coin tossing).

**11.4.4.** Prove that the expected value (average over all possibilities) of the number of visited leaves does not exceed $3^n$ for every AND-OR-tree with $2n$ levels (having $4^n$ vertices).

Let us start with $n = 1$, i.e., with a tree that has 2 levels and 4 leaves. Let the root be an AND-vertex. If the root value is 0, either both sons have 0 or one of them has 1 and the other one has 0. In the first case two leaves are enough (after finding the first zero we ignore the other son). In the second case (sons have values 0 and 1) with probability $1/2$ we start with the 0-son. Then we visit two leaves to find its value and ignore other leaves. With probability $1/2$ we start with the other (1) son and need three or four leaves. Finally, if the root has value 1, then both sons have value 1, and finding each value we use at most $3/2$ leaves in average (with probability at least $1/2$ we get 1 in the first leaf and ignore the second one). In all cases the average number of leaves does not exceed 3.

Then (for $n > 1$) we continue by induction. Assume that for any tree with $2k$ levels the average number of leaves needed by the algorithm does not exceed $3^k$. Consider a tree with $2k + 2$ levels and some fixed values in its leaves. For every choice of the order in the root and its two sons (8 possibilities) we know which vertices of level 2 (grandsons of the root) are needed. By assumption, we need (in average) at most $3^k$ leaves to determine each of these grandsons. It remains to take into account the final averaging over 8 orderings in the first two levels and note that in average we need at most 3 grandsons. □

**11.4.5.** Prove a better bound for the average number of leaves in the preceding problem.

[*Hint.* Use different bounds depending on whether the root has value 0 or 1; this replaces $\sqrt{3}$ in the bound by a smaller number $(1 + \sqrt{33})/4$.] □

**11.4.6.** Prove that for any values of tree leaves there exists an ordering which requires only $2^n$ visited leaves. □

## 11.5 A retrospective analysis

Is the improvement provided by $\alpha$-$\beta$ pruning significant (compared to the exhaustive search)? The answer is yes and no. Yes, because we get (in average) $3^n$ leaves instead of $4^n$, and the improvement factor $(4/3)^n$ grows exponentially fast as $n$ increases. No, because even $3^n$ grows exponentially fast, and the algorithm remains completely impractical even for moderate values of $n$ and relatively simple games.

So how the computer programs could play (say) chess quite well? The answer is that they play quite well (whatever this means), but not perfectly (cannot use the optimal strategy). Usually in practice some *estimate* function on the positions is introduced; it is an easily computable function that somehow reflects the advantages of the position (e.g., by counting the number of pieces of different types in chess). Then the true game is replaced by its bounded version where after some (relatively small) number of moves ($k$) the game is stopped and the outcome is determined by the estimated value of the position. In this bounded version we may apply $\alpha$-$\beta$ pruning to compute the costs (and the optimal move). Of course, this does not guarantee anything in the real game, but we cannot do anything with this.

However, there are some cases when we can find the exact value of a given position. This happens for the endgames in chess. For example, one can find exactly how many moves are needed to win if you have a king, a knight and a bishop against a lone king (knowing the exact positions of the pieces). Note that an exhaustive search is still infeasible: the number of moves needed to win could be several dozen, and each move can be done in dozens of different ways. Even an $\alpha$-$\beta$-optimized search is not an option.

The next problem shows how it can be done.

**11.5.1.** Suggest another approach that is based on the fact that the total number of possible positions is not that big.

Say for the four pieces mentioned above it is about $64^4 = 2^{24}$, i.e., about 16 million positions, or 32 million if we take into account who (White or Black) should make the next move. The array of this size easily fits into the memory of modern computers.

*Solution.* Create an array that stores an integer for each position. (The index in this array represents a position.) First we put zeros in the positions where the current player loses immediately (a mate according to the chess rules occurs). These zeros indicate that no moves are possible after that. Then we make one more pass through the array and put 1 into the positions where there is a move (allowed by the chess rules) leading to a 0-position. Being in one of these 1-positions, a player can win the game in one move. At the next pass we place $-2$ into (yet unmarked) positions such that every move leads to a 1-position. Then we place 3 into (yet unmarked) positions such that some move leads to a $(-2)$-position, then we place $-4$ into positions where every move leads to a 1- or 3-position, etc.

When no new marks appear, we stop. At that moment we know which positions are winning ones (those with positive marks) and how many moves are needed to win. $\qquad \square$

In fact this argument repeats the proof of the Zermelo theorem (also providing information about the number of moves left if both players do their best).

**11.5.2.** Could some positions be left unmarked during this procedure?

*Answer.* Yes; these are positions where both players could avoid losing the game for an arbitrarily long time. (In real chess there is a rule of threefold repetition which makes the situation much more complicated, since the full game position includes information about previous positions on the board.) $\qquad \square$

There is one chess puzzle which is especially suited to this kind of retrospective analysis (communicated by A. L. Brudno). In this puzzle the white king is placed on c3 and cannot move; White has also a queen and needs to checkmate the black king (there are no other black pieces). The immobilization of the white king makes the task more difficult for a human player, but at the same time it simplifies the analysis making the number of positions much smaller: we get about $64^2$ positions only, and the analysis can be performed even by a very primitive computer (like the first 8-bit personal computers that had only a few kilobytes of memory).

This retrospective analysis could be considered as an application of the dynamic programming technique: instead of computing the cost of a position again and again (when it appears in the different places in the game tree) we fill the costs array in a systematic way.

# 12

# Optimal coding

In this chapter we consider basic notions of coding theory. After introducing the notions of prefix and uniquely decodable codes (section 12.1) we prove the classical Kraft–McMillan inequality, a necessary and sufficient condition for the existence of a uniquely decodable code with given lengths of code words (section 12.2). Then (section 12.3) we discuss an algorithm that allows us to find an optimal uniquely decodable code for given letter frequencies. Finally, we discuss lower and upper bounds for the effectiveness of the optimal code (section 12.4).

## 12.1 Codes

We can use $n$-bit strings to encode $2^n$ objects since there are $2^n$ bit patterns of length $n$. For example, one may encode four bases G, C, A, T in DNA by four two-bit strings $00, 01, 10, 11$. There are $2^8 = 256$ bytes (8-bit sequences) and this is enough for Latin letters, digits, punctuation, etc.

In general, let $A$ be an *alphabet*, i.e., a finite set. Its elements are called *letters*, or *symbols*. A *code* for alphabet $A$ is a mapping (table) $\alpha$ that maps each symbol $a$ in $A$ into a binary string $\alpha(a)$. The word (bit string) $\alpha(a)$ is called a *code word* corresponding to $a$. The code words may have different lengths.

This definition does not assume that different symbols have different codes: all the symbols in $A$ can be encoded, say, by a string 0 or even by an empty string, but this code is evidently useless. A good code should be uniquely decodable.

The formal definition of a uniquely decodable code is as follows. Let $\alpha$ be a code for an alphabet $A$. For every $A$-word (a finite sequence of $A$-letters) $P$ we consider binary string $\alpha(P)$ that is obtained by replacing each letter by its code (the code words are concatenated without separators). The code $\alpha$ is *uniquely decodable* if different words have different encodings, i.e., if $\alpha(P) \neq \alpha(P')$ for any two different words $P \neq P'$.

**12.1.1.** Consider a code for a three-letter alphabet $\{a, b, c\}$ defined as follows: $\alpha(a) = 0$, $\alpha(b) = 01$ and $\alpha(c) = 00$. Is $\alpha$ an uniquely decodable code?

A. Shen, *Algorithms and Programming*, Springer Undergraduate Texts in Mathematics and Technology, DOI 10.1007/978-1-4419-1748-5_12, © Springer Science+Business Media, LLC 2010

*Solution.* No, since $\alpha(aa) = \alpha(c)$.                                                    □

**12.1.2.** Consider a different code for the same alphabet: $\alpha(a) = 0$, $\alpha(b) = 10$ and $\alpha(c) = 11$. Is it uniquely decodable?

*Solution.* Yes. Let us explain how one can reconstruct a word $P$ given its code $\alpha(P)$. If the first bit in $\alpha(P)$ is 0, then the first letter of $P$ is $a$. If the first bit is 1, then the first letter is $b$ or $c$. To distinguish between these two possibilities, we look at the second bit of $\alpha(P)$. Knowing the first letter of $P$, we discard it and its code, and repeat the same argument.                                                    □

This problem is a special case of a more general statement. A code $\alpha$ is called a *prefix free code* if none of the code words is a prefix of another one, i.e., if $\alpha(p)$ is not a prefix of $\alpha(q)$ (for $p \neq q$). (Sometimes prefix free codes are also called "prefix codes").

**12.1.3.** Prove that any prefix free code is uniquely decodable.

*Solution.* We can decode the input string from left to right. The first letter is determined uniquely. Indeed, if $\alpha(p)$ and $\alpha(q)$ are prefixes of the input, then one of the strings $\alpha(p)$ and $\alpha(q)$ is a prefix of the other one, and this is not possible for a prefix free code. After that we delete the encoding of the first letter and so on.     □

**12.1.4.** Construct an uniquely decodable code which is not a prefix free code.

[*Hint.* Let $\alpha(a) = 0$, $\alpha(b) = 01$, and $\alpha(c) = 11$. This code is "suffix free" (and therefore uniquely decodable), but not prefix free.]                                     □

**12.1.5.** Find (e.g., using Wikipedia) the encoding table for *Morse code* that was invented for telegraph communication and is now used (mainly) by amateur radio operators. Why can Morse code be used (it is not prefix free and even not uniquely decodable)?                                                                                 □

## 12.2 The Kraft–McMillan inequality

Why do people use code words of different lengths? The goal is to use shorter code words for more frequent letters. (This idea was used in Morse code where two frequent letters E and T are coded as "dot" and "dash".)

Assume that each letter $a$ of an alphabet $A$ has some *frequency $p(a)$*. All $p(a)$ are positive real numbers and the sum of all frequencies (for all letters) equals 1. For every code $\alpha$ we then define *average code length* (*expected code length*) as

$$E = \sum p(a)|\alpha(a)|.$$

The sum is taken over all letters $a \in A$ and $|\alpha(a)|$ stands for the length of the code word $\alpha(a)$. The motivation for this definition is straightforward: if letter $a$ has frequency $p(a)$ in a message of length $N$, then there are $Np(a)$ letters $a$ in the message and they require $Np(a)|\alpha(a)|$ bits in the encoding. The total number of bits is that $\sum Np(a)|\alpha(a)|$, i.e., $E$ bits per letter.

We arrive at the following problem: given the frequencies of letters, construct a (uniquely decidable) code with minimal average length. Theoretically one can do this by an exhaustive search (indeed, if some code word is very long, the code is not optimal, so the number of possibilities is finite; we need also some way to find whether a code is uniquely decodable). However, one can avoid exhaustive search, and in this section we explain how this can be done.

We need to better understand the obstacle that prevents the code words being short. Here is it: the lengths $n_1, \ldots, n_k$ of code words in a uniquely decodable code should satisfy the *Kraft–McMillan inequality*:

$$2^{-n_1} + 2^{-n_2} + \cdots + 2^{-n_k} \leqslant 1.$$

**12.2.1.** Check that this inequality is indeed true for our examples of uniquely decodable codes.  □

**12.2.2.** Prove that every prefix free code satisfies the Kraft–McMillan inequality.

*Solution.* Let us split $[0, 1]$ into two halves: the left one is called $T_0$, the right one is called $T_1$. Each of them again is split into two halves, and we get four intervals $T_{00}$, $T_{01}, T_{10}, T_{11}$ (e.g., $T_{01} = [1/4, 1/2]$). In this way every binary string $x$ corresponds to some interval $T_x$. The length of this interval is $1/2^n$ where $n$ is the length of $x$. If $x$ is a prefix of $y$, then $T_y$ is a part of $T_x$. If neither of the strings $x$ and $y$ is a prefix of the other one, then $T_x$ and $T_y$ are disjoint. (Indeed, reading $x$ and $y$ from left to right, we find the first bit where they disagree; at this step the corresponding intervals become disjoint.)

Now consider the intervals that correspond to the code words of a prefix free code. These intervals are disjoint subintervals of $[0, 1]$, therefore the sum of their lengths does not exceed 1, and we get the Kraft–McMillan inequality.  □

**12.2.3.** Let $n_1, \ldots, n_k$ be positive integers that satisfy the Kraft–McMillan inequality. Prove that there exists a prefix free code for a $k$-letter alphabet whose code words have lengths $n_1, \ldots, n_k$.

*Solution.* We again use the geometric representation of code words. Imagine we have some space, i.e., the interval $[0, 1]$, and this space should be allocated to $k$ users according to their requests. The user that comes with request $n_i$ wants to get an interval of length $2^{-n_i}$ that corresponds to some binary string of length $n_i$ (code word). In other words, users want to get "well aligned" intervals. The code should be prefix free, so different users should get disjoint intervals. We know that the total length of all requested intervals does not exceed 1. How can we satisfy all the requests? The answer is straightforward: we allocate space from left to right starting with longer intervals (that are more difficult to align correctly), then all the intervals would be well aligned.  □

**12.2.4.** Prove that it is possible to organize a correct *on-line* allocation: this means that we get the requests one by one, and should make an allocation before knowing the next requests. (All allocations are final.) It turns out that the Kraft–McMillan inequality remains sufficient for on-line allocation.

[*Hint*. We maintain free space as a union of aligned free intervals of pairwise different lengths. Each request is served using the least possible interval, and the rest of this interval is split into intervals of different lengths.]                    □

**12.2.5.** Prove that the Kraft–McMillan inequality is true for every uniquely decodable code (even if it is not a prefix free code).

*Solution.* There are several ways to solve this problem. We give a simple and elegant (but rather mysterious) solution. Assume that code words $P_1, \dots, P_k$ form a uniquely decodable code. We need to prove that their lengths satisfy the Kraft–McMillan inequality.

First, let us imagine that strings $P_i$ are made of letters $a$ and $b$ instead of 0 and 1 (not a big deal). Then let us consider an expression that is a (formal) sum of all code words:

$$P_1 + P_2 + \cdots + P_k.$$

This expression is a polynomial in two variables $a$ and $b$; its monomials are strings made up of $a$ and $b$ (without exponents, just the products). Now let us do a strange thing: take the $N$th power of this polynomial (where $N$ is some integer to be chosen later) and expand it (again without grouping the same variables in the monomials):

$$(P_1 + P_2 + \cdots + P_k)^N = \text{sum of monomials}.$$

For example, our code 0, 10, 11 is now written as $a, ba, bb$ and for $N = 2$ we get

$$
\begin{aligned}
(a + ba + bb)^2 &= (a + ba + bb)(a + ba + bb) \\
&= aa + aba + abb + baa + baba + babb + bba + bbba + bbbb.
\end{aligned}
$$

Note that all the monomials in the product are different as strings (of non-commuting variables), and for a good reason: no string made of $a$ and $b$ can appear twice in the sum since this would contradict the unique decoding property. (Each monomial in the sum is an encoding of some string, and different terms encode different strings.)

Now let us substitute $a = b = 1/2$: the equality is valid for any $a$ and $b$. We have

$$(2^{-n_1} + 2^{-n_2} + \cdots + 2^{-n_k})^N$$

in the left-hand side. Note that in the parentheses we have the same expression that appears in the Kraft–McMillan inequality; we need to prove that this expression does not exceed 1. The upper bound for the right-hand side can be obtained by sorting all the monomials according to their lengths. There are at most $2^l$ monomials of length $l$ (no string of length $l$ appears twice); each of them equals $1/2^l$ so in total they contribute at most 1, and the right-hand side does not exceed the maximal length of monomials, i.e., $N \max n_i$. Therefore,

$$(2^{-n_1} + 2^{-n_2} + \cdots + 2^{-n_k})^N < N \max n_i,$$

for every positive integer $N$. But if the base in the left-hand side were greater than 1, this inequality would be false for large $N$ (the exponent grows faster than the linear function).

This contradiction shows that every uniquely decodable code satisfies the Kraft–McMillan inequality.                                                                    □

## 12.3 Huffman code

Now the problem of finding the optimal code (that has minimal expected length) is reduced to the following problem: for given positive reals $p_1, \ldots, p_k$ whose sum is 1, find positive integers $n_1, \ldots, n_k$ that satisfy the Kraft–McMillan inequality and make the sum

$$\sum_{i=1}^{k} p_i n_i$$

as small as possible. Problem 12.2.5 guarantees that no uniquely decodable code can beat this minimal value, and problem 12.2.3 guarantees that an optimal code (with this average length) can be found among prefix free codes.

So how can we find these optimal $n_1, \ldots, n_k$?

**12.3.1.** Prove that for $k = 2$ (a code for a two-letter alphabet) the optimum is achieved by two 1-bit code words (independently of the frequencies $p_1$ and $p_2$).   □

Let us start the analysis of the general case with a simple observation.

**12.3.2.** Assume that the letters are numbered in the order of decreasing frequencies: $p_1 > p_2 > \ldots > p_k$. Prove then that the code words (in an optimal code) have non-decreasing lengths: $n_1 \leqslant n_2 \leqslant \ldots \leqslant n_k$.

*Solution.* If it is not the case, i.e., some letter has a shorter code than a more frequent letter, then we can decrease the average length by exchanging codes of these two letters.                                                                     □

**12.3.3.** Can we replace the assumption of decreasing frequencies by an assumption of non-increasing frequencies (several letters can have the same frequency)?

*Solution.* No. Consider an alphabet that has three equiprobable letters (i.e., $p_1 = p_2 = p_3 = 1/3$). In this case an optimal code has lengths 1, 2, 2 (it is impossible to have two code words of length 1 and some third one), and these lengths can be ordered arbitrarily.                                                                    □

Nevertheless, while searching for an optimal code (in the case of non-increasing frequencies), we may restrict our attention to the codes where the lengths form a non-decreasing sequence (indeed, the code words for equiprobable letters can be reordered without changing the expected length).

**12.3.4.** Assume that frequencies form a non-increasing sequence ($p_1 \geqslant p_2 \geqslant \ldots \geqslant p_k$), and the lengths (in an optimal code) form a non-decreasing sequence $n_1 \leqslant n_2 \leqslant \ldots \leqslant n_k$. Prove that $n_{k-1} = n_k$ (assuming that $k \geqslant 2$).

*Solution.* If it is not the case, there is only one code word of maximal length $n_k$. Then the Kraft–McMillan inequality is strict, since all the terms in the sum, except for the last term, are multiples of $2 \cdot$ (last term), and the sum cannot be equal to 1. Moreover, the "reserve" is not less than the last term. Therefore, we can decrease $n_k$ by 1 not violating the Kraft–McMillan inequality, and the code is not optimal.  □

This problem shows that we can look for an optimal code among codes where two rarest letters have codes of the same length.

**12.3.5.** Explain how this restricted problem for $k$ frequencies

$$p_1 \geqslant p_2 \geqslant \ldots \geqslant p_{k-2} \geqslant p_{k-1} \geqslant p_k$$

can be reduced to the search for an optimal code for $k - 1$ frequencies

$$p_1, p_2, \ldots, p_{k-2}, p_{k-1} + p_k$$

(we sum up the frequencies of two rarest letters).

*Solution.* We know already that we may assume without loss of generality that $n_{k-1} = n_k$. In this case the Kraft–McMillan inequality can be rewritten as

$$2^{-n_1} + 2^{-n_2} + \cdots + 2^{-n_{k-2}} + 2^{-n_{k-1}} + 2^{-n_k}$$
$$= 2^{-n_1} + 2^{-n_2} + \cdots + 2^{-n_{k-2}} + 2^{-n} \leqslant 1$$

if $n_{k-1} = n_k = n + 1$. Therefore, the numbers $n_1, \ldots, n_{k-2}, n$ satisfy the Kraft–McMillan inequality for $k - 1$ letters. Let us consider $n_1, \ldots, n_{k-2}, n$ as lengths of code words for encoding an alphabet with $k - 1$ letters having frequencies $p_1, \ldots, p_{k-2}, p_{k-1} + p_k$.

The expected lengths for these two codes (with original lengths $n_1, \ldots, n_k$ and with lengths $n_1, \ldots, n_{k-2}, n$ are closely related:

$$p_1 n_1 + \cdots + p_{k-2} n_{k-2} + p_{k-1} n_{k-1} + p_k n_k$$
$$= p_1 n_1 + \cdots + p_{k-2} n_{k-2} + (p_{k-1} + p_k) n + [p_{k-1} + p_k].$$

The last term (in square brackets) depends only on frequencies (but not lengths), so it does not matter when we look for an optimal code, and it remains to find an optimal $(k - 1)$-letter code for frequencies $p_1, \ldots, p_{k-2}, p_{k-1} + p_k$. Then we let $n_{k-1} = n_k = n + 1$ where $n$ is the length for $p_{k-1} + p_k$, and get an optimal code for the original setting.  □

This argument can be used to write a recursive program that finds the optimal code lengths (for given frequencies). The recursive call in this program is performed for a smaller alphabet size until we reach the case of two letters where the optimal code consists of 0 and 1 (i.e., $n_1 = n_2 = 1$). Then we can apply problem 12.2.3 to construct the code words for (now known) lengths. But it is much easier to combine these two goals: when $n$ is replaced by two numbers (both equal to $n + 1$), the corresponding code word $P$ can be replaced by two words $P0$ and $P1$. All other code words (for the rest of the alphabet) remain the same, and we still have a prefix free code.

The code constructed by this algorithm is called *Huffman code*. We proved that Huffman code is optimal (for given frequencies).

In the following problem we estimate the number of operations needed to construct Huffman code.

**12.3.6.** Prove that after some preprocessing of frequencies $p_1, \ldots, p_k$ that requires $O(k \log k)$ operations, we can generate code words in such a way that the number of operations needed to generate a code word is proportional to its length.

[*Hint*. Note that the required time bound for preprocessing is rather strong: indeed, only to sort $k$ numbers we need $O(k \log k)$ operations. The key idea is to save the information obtained while sorting $k$ numbers for the subsequent stages. This can be done with a priority queue: we take out two rarest letters and insert back their sum using $O(\log k)$ operations. In this way we find which two letters should be combined at each step and construct a code tree (starting from the leaves): we connect the combined letter to its "halves" by edges labeled 0 and 1. For that we need $O(1)$ operations at every step. After the tree is constructed, each code word can be traced bit by bit.] □

## 12.4 Shannon–Fano code

We now know an algorithm that constructs an optimal code (having minimal expected length) for given frequencies. However, this construction does not provide any bound for the expected length of this code. The following problems give such a bound (with gap 1 between upper and lower bounds):

**12.4.1.** Show that for any positive frequencies $p_1, \ldots, p_k$ (whose sum equals 1) there exists a code of expected length at most $H(p_1, \ldots, p_k) + 1$, where

$$H(p_1, \ldots, p_n) = p_1(-\log_2 p_1) + \cdots + p_k(-\log_2 p_k).$$

The function $H(p_1, \ldots, p_n)$ is called *Shannon entropy*.

*Solution*. Assume first that the frequencies $p_i$ are (negative) powers of 2. Then the statement is almost evident: if $p_i = 2^{-n_i}$, the numbers $n_i$ satisfy the Kraft–McMillan inequality and problem 12.2.3 gives us a prefix free code with lengths $n_1, \ldots, n_k$. The expected length for this code is $\sum p_i n_i = H(p_1, \ldots, p_n)$ and the $+1$ term is not needed.

This term becomes necessary when $\log p_i$ are not integers. In this case we take minimal $n_i$ such that $2^{-n_i} \leqslant p_i$. These $n_i$ satisfy the Kraft–McMillan inequality and exceed $-\log_2 p_i$ at most by 1, so their average exceeds Shannon entropy at most by 1, as well.                                                                    □

The explicit construction of a corresponding code is given by our solution of problem 12.2.3: we let $n_i = -\lfloor \log p_i \rfloor$ (recall that $n_i$ is the smallest integer such that $2^{-n_i} \leqslant p_i$). Assuming that $p_i$ go in non-increasing order, we may allocate code words and corresponding subintervals of $[0, 1]$ from left to right.

The code constructed in this way is called *Shannon–Fano code*.

This construction may give us a suboptimal code, but the difference in expected lengths is at most 1. Indeed, the next problem shows that any code (including the optimal one) has expected length at least $H(p_1, \dots, p_k)$.

**12.4.2.** (This problem uses some calculus) Prove that for every positive frequencies $p_1, \dots, p_k$ whose sum is 1, and for every uniquely decodable code the expected length of a code word is at least $H(p_1, \dots, p_k)$.

*Solution.* Recalling the Kraft–McMillan inequality, we have to prove that

$$2^{-n_1} + \cdots + 2^{-n_k} \leqslant 1$$

implies

$$p_1 n_1 + \cdots + p_k n_k \geqslant H(p_1, \dots, p_k)$$

for every positive integers $n_1, \dots, n_k$. In fact this is true for all positive real $n_i$, not only integers. It is convenient to rewrite this statement in terms of $q_i = 2^{-n_i}$. Then it says that for every two $k$-tuples $p_1, \dots, p_k$ and $q_1, \dots, q_k$ of positive reals that (both) have sum 1, the inequality

$$p_1(-\log q_1) + \cdots + p_k(-\log q_k) \geqslant p_1(-\log p_1) + \cdots + p_k(-\log p_k)$$

holds. In other words, we need to prove that the expression

$$p_1(-\log q_1) + \cdots + p_k(-\log q_k)$$

(considered as a function of $q_1, \dots, q_k$ such that $q_i > 0$ and $q_1 + \cdots + q_k = 1$; the values of $p_1, \dots, p_k$ are fixed) reaches its minimum when $q_i = p_i$. The domain of this function is an interior part of a simplex (a triangle for $n = 3$, a tetrahedron for $n = 4$) and the function tends to $+\infty$ as we approach the boundary (since one of $q_i$ tends to zero). Therefore the minimum is achieved in the interior point of the domain. In the minimum point the gradient vector $(-p_1/q_1, \dots, -p_n/q_n)$ is orthogonal to the hyper-plane where the function is defined, i.e., all $p_i/q_i$ are equal. Since $\sum p_i = \sum q_i = 1$, we conclude that $p_i = q_i$ for all $i$.

Another explanation: the logarithm function is concave, so for every nonnegative $\alpha_1, \dots, \alpha_k$ whose sum is 1 and for every positive $x_1, \dots, x_k$ we have

$$\log\left(\sum \alpha_i x_i\right) \geqslant \sum \alpha_i \log x_i.$$

(Jensen's inequality). Now let $\alpha_i = p_i$, $x_i = q_i/p_i$; then we have $\log 1 = 0$ in the left-hand side, and $\sum p_i \log(q_i/p_i)$ is the difference between two sides of the inequality that we want to prove.                                              □

Are the codes with different lengths of code words really useful? Of course this depends on the alphabet and letter frequencies; e.g., if all frequencies are the same, we do not get anything. In typical English written texts the frequencies are different; Shannon estimates the entropy for a 26-letter alphabet with these frequencies as 4.14 (bits per letter, so to speak). The 26-letter alphabet with equiprobable letters has entropy $\log_2 26 \approx 4.7$ (bits per letter).

The difference is not that big but becomes more significant when we take into account the lowercase/uppercase difference, punctuation, and other symbols. But the real advantage is achieved when we encode pairs of letters (or bigger blocks). Huffman code is indeed used in popular compressors like `zip` that achieve much better compression ratio for texts (and are very efficient for some other types of data).

**12.4.3.** A software company M claims that it has developed a new super-effective software for file compression: this program can achieve at least 10%-compression on every file whose length is 100 000 bytes or more (and the compression is lossless, i.e., the original file can be reconstructed from its compressed version). Prove that this claim is false. □

# 13

# Set representation. Hashing

In chapter 6 we considered several representations for sets whose elements are integers of arbitrary size. However, all those representations are rather inefficient: at least one of the operations (membership test, adding/deleting an element) runs in time proportional to the number of elements in the set. This is unacceptable in almost all practical applications.

It is possible to find a set representation where all three operations mentioned run in time $C \log n$ (in the worst case) for sets with $n$ elements. One such representation is considered in the next chapter. In this chapter, we consider another set representation that may require $n$ operations in the worst case but is very efficient in a "typical" case. The method is called *hashing*.

We consider two versions of this technique. Open addressing (section 13.1) is somehow simpler (and more efficient in terms of space), especially if we do not need deletion. Then we consider (section 13.2) hashing with lists; this version of hashing is more flexible and easier to analyze.

## 13.1 Hashing with open addressing

Suppose we want to store a set of elements of type T, where the number of elements is guaranteed to be less than n. Choose a function h that is defined on elements of type T and whose values are integers in the range 0..n-1. It is desirable that this function has different values for different elements of the set that we are trying to represent (the worst case is when all the function values are the same). This function is called a *hash function*.

Our representation uses two arrays

```
val:  array [0..n-1] of T;
used: array [0..n-1] of Boolean;
```

(we write n-1 in the type definition though it is not permitted in Pascal). The set consists of val [i] for all i such that used [i] is true. (The values val [i] are all

A. Shen, *Algorithms and Programming*, Springer Undergraduate Texts
in Mathematics and Technology, DOI 10.1007/978-1-4419-1748-5_13,
© Springer Science+Business Media, LLC 2010

different.) When possible, we store an element t at position h(t), which is considered a "natural place" for t. However, it may happen that a new element t appears whose place h(t) is already used by another element (that is, used[h(t)] is true). In this case, we search to the right looking for the first unused place and put the element t there. (Here "to the right" means that the index increases; when we reach n−1, the index wraps around.) Recall that we assume that the number of elements is always less than the number of places, therefore free places do exist.

Formally speaking, the invariant relation that we maintain is the following: For any element, the interval between its natural place and its actual place is filled completely.

This invariant makes the membership test easy. Suppose we want to check if an element t is in the set. We find the natural place for t and then go to the right until we find an empty slot or t. In the first case, the element t is not in the set (a consequence of our invariant); in the second case, the element is in the set. If it is absent, we may add it (filling the unused place found). If not, we can delete it by putting False in the corresponding cell of the used array.

**13.1.1.** The last passage has a severe error. Find it and correct it.

*Solution.* The delete operation implemented as described can destroy the invariant and create an empty position between the natural and actual positions of some element. We should be more careful. After a gap appears, we move from left to right until we find another gap or an element that is not at its natural place. If the gap appears first, we have nothing to worry about. If an element is found not at its natural place, we check whether it needs to be moved to the gap that we've created. If not, we continue our search. If yes, we move the element found to the gap. A new gap appears which we deal with in the same way.                                              □

**13.1.2.** Write the programs for membership test, adding and deleting elements.

*Solution.*

```
function is_element (t: T): Boolean;
| var i: integer;
begin
| i := h (t);
| while used [i] and (val [i] <> t) do begin
| | i := (i + 1) mod n;
| end; {not used [i] or (val [i] = t)}
| is_element := used [i] and (val [i] = t);
end;

procedure add (t: T);
| var i: integer;
begin
| i := h (t);
| while used [i] and (val [i] <> t) do begin
| | i := (i + 1) mod n;
```

```
  end; {not used [i] or (val [i] = t)}
  if not used [i] then begin
    used [i] := true;
    val [i] := t;
  end;
end;

procedure delete (t: T);
  var i, gap: integer;
begin
  i := h (t);
  while used [i] and (val [i] <> t) do begin
    i := (i + 1) mod n;
  end; {not used [i] or (val [i] = t)}
  if used [i] and (val [i] = t) then begin
    used [i] := false;
    gap := i;
    i := (i + 1) mod n;
    {gap may be filled by one of i,i+1,...}
    while used [i] do begin
      if i = h (val[i]) then begin
        {i is the natural place, nothing to do}
      end else if dist(h(val[i]),i) < dist(gap,i)
          then begin
        {gap...h(val[i])...i, nothing to do}
      end else begin
        used [gap] := true;
        val [gap] := val [i];
        used [i] := false;
        gap := i;
      end;
      i := (i + 1) mod n;
    end;
  end;
end;
```

Here dist (a,b) is the distance from a to b measured clockwise; that is,

$$\text{dist (a,b)} = (b - a + n) \bmod n.$$

(We add n, because mod works best when the dividend is positive.)  □

**13.1.3.** There are many versions of hashing. For example, when we find that the natural place (say, $i$) is occupied, we look for a free place not among $i + 1, i + 2, \ldots$, but among $r(i), r(r(i)), r(r(r(i))), \ldots$ where $r$ is some mapping of the set $\{0, \ldots, n - 1\}$ into itself. What are the possible problems?

Answer. (1) We cannot guarantee that free space will be found even if we know it exists. (2) It is not clear how to fill gaps after deleting an element. (In many practical cases, deletion is not necessary, so this approach is sometimes used. The idea is that a careful choice of the function $r$ will prevent the appearance of big clusters of occupied cells.)                                                                                □

**13.1.4.** Suppose hashing is used to store the set of all English words (say, for a spelling checker). What should we add to the data to be able to find the Russian translations of all English words?

*Solution.* The array `val` (whose elements are English words) should be extended by a parallel array `rval` of their translations: if `used[i]` is true, `rval[i]` is a translation of `val[i]`                                                                                □

In mathematical terms, here we use hashing to store functions, not sets.

## 13.2 Hashing using lists

A hash function with $k$ values is a tool that reduces the storage problem for one large set to a storage problem for $k$ small sets. Indeed, after a hash function with $k$ values is chosen, any set is split into $k$ subsets corresponding to the $k$ different values of the hash function. (Some of these subsets may be empty.) If we want to perform a membership test or an add/delete operation, we compute the hash function value and determine for which of the $k$ sets the operation should be performed.

These smaller sets may be stored conveniently using references because we know the total size of all the sets but not their individual sizes. The following problem suggests an implementation.

**13.2.1.** Suppose the values of hash function h are `1..k`. For any number j in `1..k`, consider a list of all set elements z such that `h(z) = j`. Let us store those k lists using the variables

```
Content: array [1..n] of T;
Next: array [1..n] of 1..n;
Free: 1..n;
Top: array [1..k] of 1..n;
```

in the same way as we did for k stacks of limited size (p. 87). Write the corresponding procedures. (Please note that deletion is now easier than in the open addressing case.)

*Solution.* We start with `Top[i] = 0` for all `i = 1..k`. All the positions are linked in a free list as follows: `Free = 1`; `Next[i] = i+1` for `i = 1...n-1`; `Next[n] = 0`.

```
function is_element (t: T): Boolean;
  var i: integer;
begin
```

```
  i := Top[h(t)];
  {we should search in the list starting from i}
  while (i <> 0) and (Content[i] <> t) do begin
    i := Next[i];
  end; {(i=0) or (Content [i] = t)}
  is_element := (i<>0) and (Content[i]=t);
end;

procedure add (t: T);
  var i: integer;
begin
  if not is_element (t) then begin
    i := Free;
    {Free<>0; the size limit is not reached}
    Free := Next[Free];
    Content[i]:=t;
    Next[i]:=Top[h(t)];
    Top[h(t)]:=i;
  end;
end;

procedure delete (t: T);
  var i, pred: integer;
begin
  i := Top[h(t)]; pred := 0;
  {we should search in the list starting from i;
    pred is a predecessor of i in the list
    (if exists; otherwise 0)}
  while (i <> 0) and (Content[i] <> t) do begin
    pred := i; i := Next[i];
  end; {(i=0) or (Content[i] = t)}
  if i <> 0 then begin
    {Content[i]=t, the element exists
      and should be deleted}
    if pred = 0 then begin
      {t is the first element in the list}
      Top[h(t)] := Next[i];
    end else begin
      Next[pred] := Next[i]
    end;
    {it remains to return i to the free list}
    Next[i] := Free;
    Free:=i;
  end;
end;
```

□

**13.2.2.** (Requires some probability theory) Suppose a hash function with $k$ values is used to store a set of cardinality $n$. Prove that the expected number of operations in the preceding problem does not exceed $C(1 + n/k)$, if the element t is taken at random in such a way that all values of $h(t)$ have equal probabilities $(1/k)$.

*Solution.* Let $l(i)$ be the length of the list corresponding to the hash value $i$. The number of operations does not exceed $C(1 + l(h(t)))$; the expectation does not exceed $C(1 + n/k)$, since $\sum_i l(i) = n$. □

This estimate is based on the assumption that all values of $h(t)$ have the same probability. However, for a given input distribution and a given hash function this assumption may be false, and many elements of the set may share the same value of the hash function, so large clusters appear. A method that avoids this difficulty is called *universal hashing*.

The idea is to use a family of hash functions instead of just one and to choose a function from this family at random. The hope is that any fixed set behaves well for most of the functions in the family.

Let $H$ be a family of functions. Each function maps the set $T$ into a set of cardinality $k$ (say, into $0, \ldots, k - 1$). The family $H$ is called a *universal family of hash functions* if for any two distinct elements $s, t \in T$, the probability of the event $h(s) = h(t)$ (for a random function $h \in H$) is equal to $1/k$ (the functions $h \in H$ satisfying $h(s) = h(t)$ are in proportion $1/k$ with all functions in $H$).

*Remark.* A stronger requirement may be given, namely, we may require that for any two distinct elements $s, t \in H$, the values $h(s)$ and $h(t)$ (for a randomly chosen $h$) are independent random variables uniformly distributed among $0, \ldots, k - 1$. This stronger requirement is fulfilled in the examples below.

**13.2.3.** Assume that some elements $t_1, \ldots, t_n$ are added to a set stored using a hash function $h$ from a universal family $H$. Prove that for any fixed $t_1, \ldots, t_n$ the expected number of operations (the average is taken over all $h \in H$) does not exceed $Cn(1 + n/k)$.

*Solution.* By $m_i$ we mean the number of elements among $t_1, \ldots, t_n$ with hash value $i$. (Of course, the numbers $m_0, \ldots, m_{k-1}$ depend on $h$.) The number of operations we are interested in is equal to $m_0^2 + m_1^2 + \cdots + m_{k-1}^2$ up to a constant factor. (Indeed, if $s$ elements are placed in a list, the number of operations is approximately $1 + 2 + \cdots + s \sim s^2$.) The same sum of squares may be written as the number of pairs $\langle p, q \rangle$ satisfying $h(t_p) = h(t_q)$. For any fixed $p$ and $q$ the event $h(t_p) = h(t_q)$ has probability $1/k$ (assuming that $p \neq q$). Therefore, the expected value of the corresponding term is equal to $1/k$, and the expected value of the sum is roughly $n^2/k$. More precisely, we obtain $n + n^2/k$ since we need to count terms with $p = q$. □

This problem shows that the average number of operations per element (under some probabilistic assumptions) is $C(1 + n/k)$. Here $n/k$ may be called the "average load of a hash value".

**13.2.4.** Prove a similar assertion about the arbitrary sequence of additions, deletions, and membership tests (not only additions, as in the preceding problem).

[*Hint*. Let us imagine that while performing addition, search, or deletion, the element is a person that traverses the list of its colleagues with the same hash value until it finds its twin brother (an equal element) or reaches the end of the list. (In the first case, the element disappears.) By $i$-$j$-meeting we mean the event when elements $t_i$ and $t_j$ meet each other. (It may or may not happen depending on $h$.) The total number of operations is (up to a constant factor) equal to the number of meetings plus the number of elements. When $t_i \neq t_j$, the probability of an $i$-$j$-meeting does not exceed $1/k$. It remains to count the meetings of equal elements. Let us fix some value $x \in T$ and consider all operations that refer to this value. They follow the pattern: tests, addition, tests, deletion, tests, addition, etc. The meetings occur between an added element and tested elements that follow it (up to the next deletion, and including it), therefore the total number of meetings does not exceed the number of elements equal to $x$.] $\square$

Now we give several examples of universal families. For any two finite sets $A$ and $B$, the family of all functions that map $A$ into $B$ is a universal family. However, from a practical viewpoint this family is useless since to store a random function from this family, we need an array with $\#A$ elements ($\#A$ is the cardinality of $A$). If we can afford an array of that size, we do not need hashing at all.

More practical examples of universal families may be obtained using simple algebraic techniques. By $\mathbb{Z}_p$ we denote the set of all residues modulo $p$ where $p$ is a prime number; that is, the set $\{0, 1, \ldots, p-1\}$. Arithmetic operations are performed on this set modulo $p$. A universal family is formed by all linear functionals defined on $\mathbb{Z}_p^n$ with values in $\mathbb{Z}_p$. More precisely, let $a_1, \ldots, a_n$ be arbitrary elements of $\mathbb{Z}_p$ and consider the mapping

$$h : \langle x_1, \ldots, x_n \rangle \mapsto a_1 x_1 + \cdots + a_n x_n$$

We get a family of $p^n$ mappings $\mathbb{Z}_p^n \to \mathbb{Z}_p$ indexed by $n$-tuples $\langle a_1, \ldots, a_n \rangle$.

**13.2.5.** Prove that this family is universal.

[*Hint*. Let $x$ and $y$ be distinct points of the space $\mathbb{Z}_p^n$. What is the probability of the event "a random functional $\alpha$ has the same values for $x$ and $y$"? In other words, what is the probability of the event "$\alpha(x - y) = 0$"? Note that if $u$ is a nonzero vector, all possible values of $\alpha(u)$ are equiprobable.] $\square$

In the following problem, the set $\mathbb{B} = \{0, 1\}$ is taken to be the set of residues modulo 2.

**13.2.6.** Show that the family of all linear mappings of $\mathbb{B}^n \to \mathbb{B}^m$ is universal. $\square$

The hashing idea turns out to be useful in unexpected circumstances. Here is an example called the *Bloom filter* (communicated by D.V. Varsonofiev). Suppose we want to construct a spell checker to find (most of) the typos in an English text. We do not want however, to keep a list of all correct words (in all grammatic forms). We can use the following trick. Choose some positive integer $N$ and functions $f_1, \ldots, f_k$ that map character strings to $1, \ldots, N$. Consider an array of $N$ bits initially set to zero.

Then for any (correctly spelled) word $x$, compute the values $f_1(x), \ldots, f_k(x)$ and make the corresponding bits equal to 1. (Some bits may correspond to several words.) Then the approximate test to check whether a string $z$ is a correctly spelled word, is as follows. Compute all values $f_1(z), \ldots, f_k(z)$ and check that all the corresponding bits are 1s. This test may miss some errors, but all correct words will be allowed.

# 14

# Sets, trees, and balanced trees

In this chapter we consider another class of set representations that can be used if elements of our sets are taken from some ordered set. The set elements are considered as labels, and ordering is consistent with the tree structure (see section 14.1 for details). This can be useful in some applications, but if we want to have upper bounds for the number of operations in the worst case, we need to keep some kind of balance between left and right subtrees of the same vertex. One of the possible definitions of balance and corresponding balancing algorithms are considered in section 14.2.

## 14.1 Set representation using trees

### Full binary trees and $T$-trees

Draw a point. Now draw two arrows going up-left and up-right to two other points. From those two points also draw two arrows, etc. The resulting tree is called a *full binary tree* (the $n$-th level has $2^{n-1}$ points). The initial point (at the bottom of the tree) is called the *root*. Each vertex has two *sons* (arrows point to them), the *left* son and the *right* son. Each vertex (except for the root) has an unique *father*.

Please note that many textbooks draw trees with the root at the top and also use the words "child" ("parent", "sibling", etc.) instead of "son" ("father", "brother", etc.).

Now choose some subset of the set of all vertices of the full binary tree. It should satisfy the following requirement: for each vertex of the subset, its father belongs to the subset, too. (Therefore, all vertices on a path from the root to some vertex from the subset belong to the subset.) Assume that each vertex in the subset has a label that is an element of some set $T$. (In other words, we assume that a mapping from the subset into the set $T$ is given.) Such a subset with labels from $T$ is called a $T$-*tree*. The set of all $T$-trees is denoted by $\mathrm{Tree}(T)$.

The notion of $T$-tree may be defined recursively. Any nonempty $T$-tree is divided into three parts: the root (which carries a label from $T$), the left subtree, and the right subtree (one or both of which may be empty). Therefore, there is an one-to-one

A. Shen, *Algorithms and Programming*, Springer Undergraduate Texts in Mathematics and Technology, DOI 10.1007/978-1-4419-1748-5_14, © Springer Science+Business Media, LLC 2010

correspondence between the set of nonempty $T$-trees and the product $T \times \mathrm{Tree}(T) \times \mathrm{Tree}(T)$. We get the following equality:

$$\mathrm{Tree}(T) = \{\mathbf{empty}\} + T \times \mathrm{Tree}(T) \times \mathrm{Tree}(T).$$

(here **empty** stands for the empty tree).

### Subtrees and height

Assume that some $T$-tree is fixed. For any vertex $x$, the following objects are defined: the *left subtree* (the left son of $x$ and all its descendants); the *right subtree* (the right son of $x$ and all its descendants); and the *subtree rooted at $x$* (the vertex $x$ and all its descendants). The left and right subtrees of $x$ may be empty, but the subtree rooted at $x$ may not (it always contains the vertex $x$). The *height* of a subtree is defined as the maximal length of the sequence $y_1, \ldots, y_n$ of its vertices where $y_{i+1}$ is a son of $y_i$ for all $i$, minus one. (The height of the empty tree is $-1$ by definition; the height of a tree containing only the root is 0.)

### Ordered $T$-trees

Assume that a linear order is defined on the set $T$. A $T$-tree is *ordered* if the following requirement is fulfilled: for any vertex $x$, all labels in its left subtree are less than the label at $x$ and all labels in its right subtree are greater than the label at $x$.

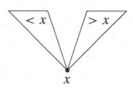

**14.1.1.**  Prove that all labels in an ordered subtree are different.

[*Hint*. Induction over the height of the tree.]                                    □

### Set representation using trees

Consider any tree as a representation of the set of labels of its vertices. (Of course, the same set may have different representations.)

If the tree is ordered, each element can easily find its way to a place in the tree. It starts from the root; coming to a vertex, an element compares itself with the label at that vertex and decides whether to go to the left or to the right.

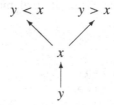

Using this rule, the element either finds the identical label already present in the tree or the place where it should stay to keep the tree ordered.

In this chapter we assume that the set $T$ is a linearly ordered set. All $T$-trees that we consider are ordered.

### Tree representation

The simplest way to represent a tree is to identify the vertices of a full binary tree with integers $1, 2, 3, \ldots$ (the left son of $n$ is $2n$, the right son of $n$ is $2n + 1$) and store the labels in an array val [1..N] (for a large enough N). However, this approach (used in section 4.2, heap sort algorithm) now wastes space because space is set aside for positions in the full binary tree that are not filled in a specific $T$-tree.

The following approach (discussed in section 7.2) is more space efficient. We use three arrays

```
val: array [1..n] of T;
left, right: array [1..n] of 0..n;
```

(n is the maximal possible number of tree vertices for trees we want to store) and a variable root: 0..n. Each vertex of the stored $T$-tree will have a number that is an integer in 1..n. Different vertices have different numbers; some numbers may be unused. The label of the vertex with number x is stored in val[x]. The root has number root. If vertex i has sons, their numbers are left[i] and right[i]. Nonexistent sons are replaced by the number 0. Similarly, the condition root = 0 means that the tree is empty.

The tree vertices only occupy part of the array. For "free" values of i that are not used as vertex numbers, the values val[i] have no meaning. We want the free numbers to be "linked in a list"; the first free number is stored in a special variable free: 0..n, while the free number that follows i in the list is left[i]. In other words, the list of all free numbers is

```
free, left[free], left[left[free]],...
```

For the last free number i in the list, the value left[i] equals 0. If free = 0, there are no free numbers. (We use the array left to link all free numbers in a list but of course, we may use the array right instead.)

We can use any other integer outside 1..n to indicate the absence of a vertex (instead of 0). To stress this, we use a symbolic constant null = 0 instead of the numeral 0.

**14.1.2.** Write a procedure that checks if an element t:T is present in an ordered tree (as described above).

*Solution.*

```
if root = null then begin
| ..t is not in the tree
end else begin
  x := root;
  {invariant: it remains to check if t is present in
    a nonempty subtree rooted at x}
  while ((t < val [x]) and (left [x] <> null)) or
        ((t > val [x]) and (right [x] <> null)) do begin
    if t < val [x] then begin {left [x] <> null}
    | x := left [x];
    end else begin {t > val [x], right [x] <> null}
    | x := right [x];
    end;
  end;
  {either t = val [x] or t is not in the tree}
  ..answer is (t = val [x])
end;                                                        □
```

**14.1.3.** Simplify the procedure using the following trick. Extend the array val, adding a cell with index null. To simplify the search, you may put t in val [null].

*Solution.*

```
val [null] := t;
x := root;
while t <> val [x] do begin
  if t < val [x] then begin
  | x := left [x];
  end else begin
  | x := right [x];
  end;
end;
..answer is (x <> null).                                    □
```

**14.1.4.** Write a procedure that adds an element t to a set represented as an (ordered) *T*-tree. (If t is already present, nothing should be done.)

*Solution.* The procedure get_free (var i:integer) produces a free integer i in 1..n (that is, an integer that is not a number of any vertex) and updates the free list. (For simplicity, we assume that free integers exist.)

```
procedure get_free (var i: integer);
begin
  {free <> null}
  i := free;
  free := left [free];
end;
```

Using this procedure, we write:

```
if root = null then begin
  get_free (root);
  left [root] := null; right [root] := null;
  val [root] := t;
end else begin
  x := root;
  {invariant: it remains to add t to a (nonempty) subtree
   rooted at x}
  while ((t < val [x]) and (left [x] <> null)) or
        ((t > val [x]) and (right [x] <> null)) do begin
    if t < val [x] then begin
      x := left [x];
    end else begin {t > val [x]}
      x := right [x];
    end;
  end;
  if t <> val [x] then begin {t is not in the tree}
    get_free (i);
    left [i] := null; right [i] := null;
    val [i] := t;
    if t < val [x] then begin
      left [x] := i;
    end else begin {t > val [x]}
      right [x] := i;
    end;
  end;
end;
```
□

**14.1.5.** Write a procedure that deletes an element t from a set represented as an ordered tree. (If the element is not in the set, nothing should be done.)

*Solution.*

```
if root = null then begin
  {the tree is empty, there is nothing to do}
```

```
end else begin
  x := root;
  {it remains to delete t from the subtree rooted at x;
   since it may require changes in the father node,
   we introduce the variables  father: 1..n and
   direction: (l, r) with the following
   invariant: if x is not the root, then father
   is (the number of) x's father node, direction is
   equal to l/r if x is the left/right son of its father}
  while ((t < val [x]) and (left [x] <> null)) or
        ((t > val [x]) and (right [x] <> null)) do begin
    if t < val [x] then begin
      father := x; direction := l;
      x := left [x];
    end else begin {t > val [x]}
      father := x; direction := r;
      x := right [x];
    end;
  end;
  {t = val [x] or t is not in the tree}
  if t = val [x] then begin
    ..delete the node x with a known father and direction
  end;
end;
```

The deletion of a vertex uses the procedure

```
procedure make_free (i: integer);
begin
  left [i] := free;
  free := i;
end;
```

which adds the number i to the free list. While deleting a vertex, we should distinguish between four cases depending on whether the vertex has a left/right son or not.

```
if (left [x] = null) and (right [x] = null) then begin
  {x is a leaf, no sons}
  make_free (x);
  if x = root then begin
    root := null;
  end else if direction = l then begin
    left [father] := null;
  end else begin {direction = r}
    right [father] := null;
  end;
```

```
    end else if (left[x]=null) and (right[x] <> null)
    |   then begin
    |   {when x is deleted, right[x] occupies its place}
    |   make_free (x);
    |   if x = root then begin
    |   |  root := right [x];
    |   end else if direction = l then begin
    |   |  left [father] := right [x];
    |   end else begin {direction = r}
    |   |  right [father] := right [x];
    |   end;
    end else if (left[x] <> null) and (right[x]=null)
    |   then begin
    |   ..the symmetrical code
    end else begin {left [x] <> null, right [x] <> null}
    |   ..delete a vertex with two sons
    end;
```

The deletion of a vertex with two sons is the most difficult case. Here we should exchange it with a vertex that contains the next label (in the sense of label ordering).

```
    y := right [x]; father := x; direction := r;
    {now father and direction refer to vertex y}
    while left [y] <> null do begin
    |   father := y; direction := l;
    |   y := left [y];
    end;
    {val[y] is minimal element of the set larger
     than val[x], y has no left son}
    val [x] := val [y];
    ..delete the vertex y (we already know how to do
       that for a vertex without the left son)                     □
```

**14.1.6.** Simplify the deletion procedure using the following observation: Some cases (say, the first two) may be combined into a single case.                     □

**14.1.7.** Use an ordered tree to store a function whose domain is a finite subset of $T$ and whose range is some set $U$. The operations are: find the value of the function for a given argument; change this value; delete an element from the domain; and add an element to the domain (the value is also provided).

*Solution.* We represent the domain using an ordered tree and add one more array

```
    func_val: array [1..n] of U;
```

If val[x] = t and func_val[x] = u, then the function value on t equals u.                     □

**14.1.8.** Assume that we want to find the $k$-th element of a set (according to the ordering on $T$) in time limited by $C \cdot$ (tree height) plus some constant. What additional information do we need to store at the tree vertices?

*Solution.* At each vertex, we store the number of its descendants. When a vertex is added or deleted, this information must be updated along a path from the root to the new/deleted vertex. While searching for the $k$-th vertex, we maintain the following invariant: the vertex in question is the $s$-th vertex (according to the $T$-ordering) of a subtree rooted at $x$ (here $s$ and $x$ are variables). $\qquad\square$

### Running time

All of the procedures discussed above (membership test, addition, and deletion) run in time $C \cdot$ (tree height) plus constant. (The constant is needed for an empty tree or a one-element tree where the height is $-1$ or $0$.) For a "well-balanced" tree where all leaves are approximately at the same height, the tree height is close to the logarithm of the number of vertices. However, for a unbalanced tree the height may be much larger. In the worst case, the vertices may form a chain (if all vertices have no left son, for example) and the tree height is the number of vertices (minus one). This happens if we start with the empty set and add elements in increasing order. However, one can prove that if the elements are added in random order, then the expected height of the tree will not exceed $C \log$(tree size). If this "average bound" is not good enough for our application, we must expend additional effort to keep the tree "balanced". This is explained in the next section.

## 14.2 Balanced trees

A tree is called *balanced* (or an AVL-*tree*, in honor of the inventors of this algorithm, G.M. Adelson-Velsky and E.M. Landis) if for any vertex, the heights of the left and the right subtrees differ by at most 1. (In particular, the only son of a vertex is required to be a leaf since the height of the other subtree is $-1$.)

**14.2.1.** Find the minimal and maximal possible number of vertices in a balanced tree of height $n$.

*Solution.* The maximal number of vertices is equal to $2^{n+1} - 1$. If $m_n$ is the minimal number of vertices, then $m_{n+2} = 1 + m_n + m_{n+1}$. An easy induction argument gives $m_n = \Phi_{n+3} - 1$ (where $\Phi_n$ is the $n$-th Fibonacci number: $\Phi_1 = 1$, $\Phi_2 = 1$, and $\Phi_{n+2} = \Phi_n + \Phi_{n+1}$). $\qquad\square$

**14.2.2.** Prove that a balanced tree with $n > 1$ vertices has height at most $C \log n$ for some constant $C$ that does not depend on $n$.

*Solution.* By induction over $n$, we prove that $\Phi_{n+2} \geqslant a^n$ where $a$ is the larger root of the quadratic equation $a^2 = 1 + a$; that is, $a = (\sqrt{5} + 1)/2$. (This number is usually called "the golden mean".) It remains to apply the preceding problem. $\qquad\square$

## Rotations

After an element is added or deleted, the tree may become unbalanced, and we have to restore the balance. Therefore, we need some tree transformations that preserve the set of labels and the ordering requirement, but help to balance the tree. Here are some of those transformations:

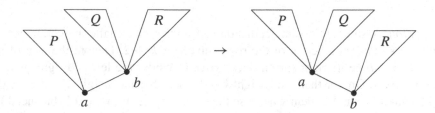

Assume that a vertex $a$ has a right son $b$. Let $P$ be the left subtree of $a$. Let $Q$ and $R$ be the left and right subtrees of $b$, respectively. The ordering requirement guarantees that $P < a < Q < b < R$. (This means that any label in $P$ is smaller than $a$, that $a$ is smaller than any label in $Q$, etc.) The same condition is imposed by the ordering requirements for another tree. The latter tree has root $b$; the left son $a$ of the root has left subtree $P$ and right subtree $Q$; the right subtree of the root is $R$. Therefore the first tree may be transformed to the second one without changing the set of labels or violating the ordering requirements. This transformation is called a *small right rotation*. It is called "right" because there is a symmetric "left" rotation; it is called "small" because there exists a "big" rotation which we describe now.

Let $b$ be the right son of the root vertex $a$; let $c$ be the left son of $b$; let $P$ be the left subtree of $a$; let $Q$ and $R$ be the left and the right subtrees of $c$, respectively; and finally, let $S$ be the right subtree of $b$. Then $P < a < Q < c < R < b < S$.

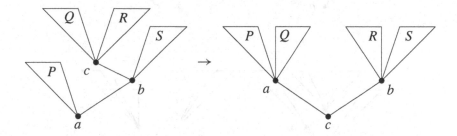

The same ordering conditions are imposed by a tree with root $c$; its left son $a$ and right son $b$ that have the left and the right subtrees $P$ and $Q$ (for $a$) and $R$ and $S$ (for $b$). The corresponding transformation is called a *big right rotation*. (A *big left rotation* is defined in a symmetric way.)

**How to balance a tree using rotations**

**14.2.3.** Suppose a tree is balanced everywhere except at the root where the difference of heights between the left and right subtrees equals 2 (that is, the left and right subtrees are balanced and their heights differs by 2). Prove that this tree may be transformed into a balanced tree using one of the four transformations mentioned above and that the height remains the same or decreases by 1 after the transformation.

*Solution.* Assume, for example, that the left subtree has smaller height, which we denote by $k$. Then the height of the right subtree is $k + 2$. Denote the root of the tree by $a$. Let $b$ be its right son (it does exist). Consider the left and right subtrees of the vertex $b$. One of them has height $k + 1$, the other has height $k$ or $k + 1$. (Its height cannot be smaller than $k$ because the right subtree of the root is balanced.) If the height of the left subtree of $b$ is $k + 1$, and the height of the right subtree of $b$ is $k$, a big right rotation is needed; in all other cases, a small right rotation suffices. Here are the three possible cases:

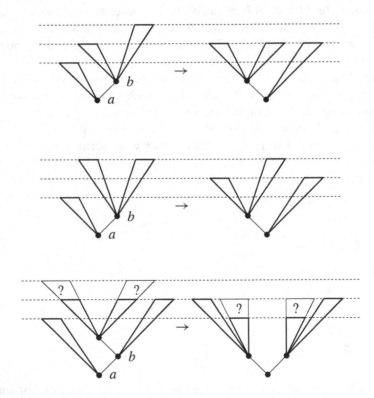

**14.2.4.** A leaf is added to or deleted from a balanced tree. Prove that it is possible to make the tree balanced again using several rotations and that the number of rotations does not exceed the tree height.

*Solution.* We prove the more general statement:

*Lemma.* If a subtree $Y$ of a balanced tree $X$ is replaced by a balanced tree $Z$, and the heights of $Y$ and $Z$ differ by 1, then the resulting tree can be made balanced by several rotations. The number of rotations does not exceed the height where the change occurs (that is, where the root of $Y$ and $Z$ is located).

The addition/deletion of a leaf is a special case of the transformation mentioned in the lemma, therefore it is enough to prove this lemma.

*Proof* of the lemma. We use induction over the height where the change is made. If the change is made at the root, the entire tree is replaced; in this case, the lemma is evident because the tree $Z$ is balanced. Assume that the replaced tree $Y$ is, say, the left subtree of some vertex $x$. Two cases are possible:

1. After replacement, the balance condition at the vertex $x$ is still valid. (However, the balance condition at the ancestors of $x$ may be violated because the height of the subtree rooted at $x$ may change.) In this case, we apply the induction hypothesis assuming that the replacement was done at the lower level and the whole tree rooted at $x$ was replaced.

2. The balance condition at $x$ is no longer valid. In this case, the height difference is 2 (it cannot be larger because the heights of $Y$ and $Z$ differ by at most 1). Here two subcases are possible:

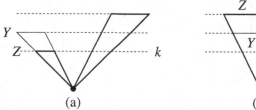 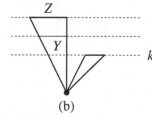

(a)                                         (b)

a) The right subtree of $x$ (the one that was not replaced) is higher. Assume that the height of the left subtree of $x$ (i.e., $Z$) is $k$; then the height of the right subtree is $k + 2$. The height of the old left subtree of $X$ (i.e., $Y$) was $k + 1$. The subtree of the initial tree rooted at $x$ has height $k + 3$ and its height does not change after replacement.

   By the preceding problem, a rotation can transform the subtree rooted at $x$ into a balanced subtree of height $k+2$ or $k+3$. While doing this, the height of the subtree rooted at $x$ (compared with its height before the transformation) did not change or was decreased by 1. Therefore, we apply the induction assumption.

b) The left subtree of $x$ is higher. Let the height of the left subtree (i.e., $Z$) be $k + 2$; the right subtree has height $k$. The old left subtree of $x$ (i.e., $Y$) was of height $k + 1$. The subtree rooted at $x$ (in the initial tree) has height $k + 2$; after the replacement it has height $k + 3$. After a suitable rotation (see the preceding problem), the subtree rooted at $x$ becomes balanced and its height is $k+2$ or $k+3$; therefore, the change in height (compared with the height of the subtree of $X$ rooted at $x$) does not exceed 1 and the induction assumption applies.                                                                                □

**14.2.5.** Write addition and deletion procedures that keep the tree balanced. The running time should not exceed $C \cdot$ (tree height). It is allowed to store additional information (needed for balancing) at the vertices of the tree.

*Solution.* For each vertex we keep the difference between the heights of its right and left subtrees:

$$\mathtt{diff}\ [\mathtt{i}] = \text{(the height of the right subtree of i)}$$
$$- \text{(the height of the left subtree of i)}.$$

We need four procedures that correspond to left/right, small/big rotations. Let us first make two remarks. (1) We want to keep the number of the tree root unchanged during the rotation. (Otherwise it would be necessary to update the pointer at the father vertex, which is inconvenient.) This can be done, because the numbers of tree vertices may be chosen independently of their content. (In our pictures, the number is drawn near the vertex while the content is drawn inside it.)

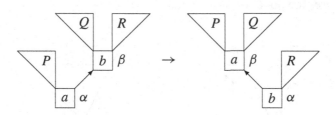

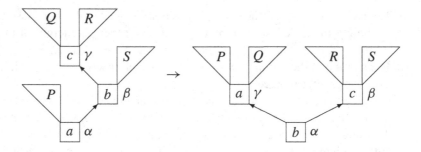

(2) After the transformation, we should update values in the $\mathtt{diff}$ array. To do this, it is enough to know the heights of trees $P$, $Q$, ... up to a constant (only differences are important), so we may assume that one of the heights is equal to 0.

Here are the rotation procedures:

```
procedure SR (a:integer); {small right rotation at a}
  var b: 1..n; val_a,val_b: T; h_P,h_Q,h_R: integer;
```

```
begin
  b := right [a]; {b <> null}
  val_a := val [a]; val_b := val [b];
  h_Q := 0; h_R := diff[b];
  h_P := (max(h_Q,h_R)+1)-diff[a];
  val [a] := val_b; val [b] := val_a;
  right [a] := right [b] {subtree R}
  right [b] := left [b] {subtree Q}
  left [b] := left [a] {subtree P}
  left [a] := b;
  diff [b] := h_Q - h_P;
  diff [a] := h_R - (max (h_P, h_Q) + 1);
end;

procedure BR(a:integer);{big right rotation at a}
  var b,c: 1..n; val_a,val_b,val_c: T;
      h_P,h_Q,h_R,h_S: integer;
begin
  b := right [a]; c := left [b]; {b,c <> null}
  val_a := val [a]; val_b := val [b]; val_c := val [c];
  h_Q := 0; h_R := diff[c];
  h_S := (max(h_Q,h_R)+1)+diff[b];
  h_P := 1 + max (h_S, h_S-diff[b]) - diff [a];
  val [a] := val_c; val [c] := val_a;
  left [b] := right [c] {subtree R}
  right [c] := left [c] {subtree Q}
  left [c] := left [a]  {subtree P}
  left [a] := c;
  diff [b] := h_S - h_R;
  diff [c] := h_Q - h_P;
  diff [a] := max (h_S, h_R) - max (h_P, h_Q);
end;
```

The (small and big) left rotations are similar.                        □

The addition/deletion procedures are written as before, but now they have to update the diff array and restructure the tree to keep it balanced.

An auxiliary procedure with the following pre- and postconditions is used:

**before:** the left and right subtrees of the vertex number a are balanced; the difference of heights at a is at most 2; the diff array is filled correctly for the subtree rooted at a;

**after:** the subtree rooted at a is now balanced; the diff is updated (inside that subtree); the change in the height of the subtree rooted at a is stored in d and is equal to 0 or -1; the remaining part of the tree (including the diff array) remains unchanged.

```
procedure balance (a: integer; var d: integer);
begin {-2 <= diff[a] <= 2}
  if diff [a] = 2 then begin
    b := right [a];
    if diff [b] = -1 then begin
    | BR (a); d := -1;
    end else if diff [b] = 0 then begin
    | SR (a); d := 0;
    end else begin {diff [b] = 1}
    | SR (a); d := - 1;
    end;
  end else if diff [a] = -2 then begin
    b := left [a];
    if diff [b] = 1 then begin
    | BL (a); d := -1;
    end else if diff [b] = 0 then begin
    | SL (a); d := 0;
    end else begin {diff [b] = -1}
    | SL (a); d := - 1;
    end;
  end else begin {-2<diff[a]<2, there is nothing to do}
  | d := 0;
  end;
end;
```

To restore the balance, we go downwards from a leaf to the root. To do that, we store the path from the root to the current vertex in a stack. The elements of the stack are pairs ⟨vertex, direction of move from the vertex⟩; that is, values of type

```
record
| vert: 1..n; {vertex}
| direction : (l, r); {l for left, r for right}
end;
```

The addition of an element t is now as follows:

```
if root = null then begin
| get_free (root);
| left[root]:=null; right[root]:=null; diff[root]:=0;
| val[root]:=t;
end else begin
| x := root;
| ..make the stack empty
| {invariant: it remains to add t to the nonempty subtree
|   rooted at x; the stack contains the path to x}
| while ((t < val [x]) and (left [x] <> null)) or
|       ((t > val [x]) and (right [x] <> null)) do begin
```

```
    if t < val [x] then begin
    | ..add <x, l> to the stack
    | x := left [x];
    end else begin {t > val [x]}
    | ..add <x, r> to the stack
    | x := right [x];
    end;
  end;
  if t <> val [x] then begin {t is not in the tree}
    get_free (i); val [i] := t;
    left [i] := null; right [i] := null; diff [i] := 0;
    if t < val [x] then begin
    | ..add <x, l> to the stack
    | left [x] := i;
    end else begin {t > val [x]}
    | ..add <x, r> to the stack
    | right [x] := i;
    end;
    d := 1;
    {invariant: the stack contains the path to a changed
     subtree whose height has increased by d (= 0 or 1);
     this subtree is balanced; values of diff for its
     vertices are correct; in the remaining part of the
     tree everything is unchanged (including the values
     of diff)}
    while (d <> 0) and the stack is nonempty do begin
      {d = 1}
      ..take a pair from stack into <v, direct>
      if direct = l then begin
      | if diff [v] = 1 then begin
      | | c := 0;
      | end else begin
      | | c := 1;
      | end;
      | diff [v] := diff [v] - 1;
      end else begin {direct = r}
      | if diff [v] = -1 then begin
      | | c := 0;
      | end else begin
      | | c := 1;
      | end;
      | diff [v] := diff [v] + 1;
      end;
      {c = the change in the height of the subtree rooted
       at v (compared with the initial tree); the array
```

```
          diff has correct values inside that subtree; the
          balance condition at v may be violated}
         balance (v, d1); d := c + d1;
       end;
     end;
  end;
```

It is easy to check that d may be equal to 0 or 1 (but not −1); indeed, if c = 0, then diff [v] = 0 and balancing is not performed.

The deletion procedure is similar. Its main part is:

```
{invariant: the stack contains a path to the changed
  subtree whose height was changed by d (=0 or -1)
  compared with the initial tree; this subtree is
  balanced; the values of diff are correct for the
  vertices of that subtree; the remaining part of the
  tree is unchanged (including the values of diff)}
while (d <> 0) and the stack is not empty do begin
  {d = -1}
  ..take a pair from the stack into <v, direct>
  if direct = 1 then begin
    if diff [v] = -1 then begin
    | c := -1;
    end else begin
    | c := 0;
    end;
    diff [v] := diff [v] + 1;
  end else begin {direct = r}
    if diff [v] = 1 then begin
    | c := -1;
    end else begin
    | c := 0;
    end;
    diff [v] := diff [v] - 1;
  end;
  {c = the change in the height of the subtree rooted
   at v (compared with the initial tree); the array diff
   has correct values inside that subtree; the balance
   condition at v may be violated}
  balance (v, d1);
  d := c + d1;
end;
```

It is easy to check that d may be equal to 0 or −1 (but not −2); indeed, if c = −1, then diff [v] = 0 and balancing is not performed.

Let us mention that the existence of the stack makes the variables `father` and `direction` used in the deletion procedure (see above) redundant, because now the stack top contains the same information.

**14.2.6.** Prove that while the element is added,

(a) the second case of the balancing step (see picture on p. 212) is in fact, impossible;

(b) the complete balancing of the entire tree requires only one rotation.

However, deletion may require many rotations to restore the balance. □

*Remark.* Addition and deletion procedures may be simplified if we do not want to make them similar.

### Other versions of balanced trees

There are several other ways to represent sets using trees. Some of those methods also guarantee a running time of order $\log n$ for each operation. Let us sketch one of them, called *B-trees*. (It is often used for large databases stored on a hard disk.)

Up to now each vertex contained only one element of the set. This element was used as a threshold that separates the left and right subtrees. Now let the vertex store $k \geqslant 1$ elements of the set. The value of $k$ may be different for different vertices and may change while adding or deleting elements (see below). The $k$ elements stored at a vertex are used as separators between $k + 1$ subtrees (so a vertex with $k$ elements may have up to $k + 1$ sons).

Assume that some number $t \geqslant 1$ is fixed. We consider trees that satisfy the following requirements:

1. Each vertex contains not less than $t$ and not more than $2t$ elements. (The root is an exception; it may contain any number of elements not exceeding $2t$.)
2. Any vertex with $k$ elements either has $k + 1$ sons or does not have any sons at all (that is, it is a *leaf*).
3. All leaves are on the same level.

The *addition* of an element proceeds as follows. If the leaf where this element goes is not full (that is, contains less than $2t$ elements), we simply add this element to that leaf. If that leaf is full, then we have $2t + 1$ elements ($2t$ old ones and the new one). We split them into two leaves with $t$ elements and the median element between them. This median element should be added to a vertex at the preceding level. This is easy if that vertex has less than $2t$ elements. If it is full, then it is split into two vertices, a median is found, etc. Finally, if we need to add the new element to the root and the root is full, we split the root into two vertices and the tree height is increased by 1.

The *deletion* of an element that is placed not at a leaf may be reduced to the deletion of the next element of the set, which is in a leaf. Therefore, it is enough to delete elements from leaves. If the leaf becomes too small, we can borrow some elements from a neighboring leaf, unless it too has the minimal possible size $t$. If both leaves have size $t$, together they have $2t$ elements, or rather $2t + 1$ elements if we count the separator between them. After deleting one element, the remaining $2t$ elements may be placed onto one leaf. However, the vertex of the preceding level may now be too small. In that case, we have to do the same transformation at that level, etc.

**14.2.7.** Implement this scheme of set representation and check that it also performs additions, deletions, and membership tests in time $C \log n$, where $n$ is the cardinality of the set.                                                                    □

**14.2.8.** Another definition of a balanced tree requires that for each vertex the number of vertices in its left and right subtrees do not differ too much. (The advantage of this definition is that a rotation performed at some vertex does not destroy the balance at the ancestors of that vertex.) Using this idea, find a set representation that guarantees a running time bound of $C \log n$ for additions, deletions and membership tests.

[*Hint*. This approach also uses small and big rotations. The details can be found in the book of Reingold, Nievergelt, and Deo [11].]                              □

# 15

# Context-free grammars

In chapter 5 we use finite automata for text parsing. As noted, there are rather simple structures (e.g., nested comments) that cannot be parsed with finite automata. There is a more powerful formalism called context-free grammars that is often used when finite automata are not enough. In section 15.1 we define context-free grammars and consider a general polynomial parsing algorithm. However, this algorithm is not fast enough to be practical, and in the next two sections we consider faster (linear time) algorithms that can be used for some classes of context-free grammars, called recursive-descent parsing (section 15.2) and LL-parsing (section 15.3).

## 15.1 General parsing algorithm

To define a *context-free grammar* we should:

- fix a finite set $A$, called an *alphabet*, whose elements are called *symbols* or *letters*; finite sequences of symbols are called *strings* or *words*;
- divide all symbols in $A$ into two classes: *terminal* symbols and *nonterminal* symbols;
- choose a nonterminal symbol called the *initial symbol*, or *axiom*;
- fix a finite set of *productions* or *production rules*; each production has the form $K \to X$, where $K$ is some nonterminal and $X$ is a string that may contain both terminal and nonterminal symbols.

The name "context-free" is used since a production rule $K \to X$ can be applied wherever we see $K$, the context (letters around $K$) does not matter. We often omit the words "context-free" because we do not consider other types of grammars.

Assume that a context-free grammar is given. A *derivation* in this grammar is a sequence of strings $A_0, A_1, \ldots, A_n$, where $A_0$ is a one-letter string consisting of the initial symbol; $A_{i+1}$ is obtained from $A_i$ by replacing some nonterminal $K$ in $A_i$ by a string $X$ according to one of the production rules $K \to X$.

A string containing only terminals is *generated* by a grammar if there exists a derivation that ends in this string. The set of all strings generated by some grammar $G$

A. Shen, *Algorithms and Programming*, Springer Undergraduate Texts in Mathematics and Technology, DOI 10.1007/978-1-4419-1748-5_15,

is called the *context-free language generated by G*. A language (that is, a set of strings) is called *context-free* if it is generated by some context-free grammar.

In this chapter, as well as the following one, we are interested in the following question: A context-free grammar *G* is given; construct an algorithm that checks if an input string belongs to the language generated by *G*.

*Example 1*. Alphabet:

$$( \ ) \ [ \ ] \ E$$

(four terminals and one nonterminal E). Axiom: E. Productions:

$$E \to (E)$$
$$E \to [E]$$
$$E \to EE$$
$$E \to$$

(the last rule has the empty string on its right-hand side).

Examples of generated strings:

$$(\text{empty string})$$
$$()$$
$$([])$$
$$()[([])]$$
$$[()[]()[]]$$

Examples of strings not in the language:

$$($$
$$)($$
$$(]$$
$$([)]$$

This grammar was considered in chapter 6. An algorithm that checks whether an input string belongs to the corresponding language was considered; that algorithm used a stack.

*Example 2*. Another grammar that generates the same language:

Alphabet: ( ) [ ] T E

Productions:

$$E \to$$
$$E \to TE$$
$$T \to (E)$$
$$T \to [E]$$

In all subsequent examples, the axiom will be the nonterminal on the left-hand side of the first rule unless stated otherwise (in this example, the axiom is E).

For any nonterminal $K$, consider the set of all strings composed of terminals that can be obtained from $K$ by a derivation. (For the axiom, this set is a language generated by a grammar.) In a sense, each rule of the grammar is a statement about those sets. Let us explain what we mean using the grammar of example 2. Let $T$ and $E$ be the sets of all strings in the alphabet $\{\,(\,,\,)\,,\,[\,,\,]\,\}$ derivable from nonterminals T and E, respectively. The rules of the grammar (left column) correspond to the following properties of $T$ and $E$ (right column).

| | |
|---|---|
| E → | $E$ contains an empty string |
| E → TE | if $A$ is in $T$ and $B$ is in $E$, then $AB$ is in $E$ |
| T → [E] | if $A$ is in $E$, then $[A]$ is in $T$ |
| T → (E) | if $A$ is in $E$, then $(A)$ is in $T$ |

These four properties of $E$ and $T$ do not determine those sets uniquely. For example, they are still true if $T = E =$ the set of all strings. However, one may prove (for an arbitrary context-free grammar) that the sets defined by the grammar are minimal among all the sets having those properties ("minimal" means "minimal up to inclusion").

**15.1.1.** Give the precise statement and proof of this claim. □

**15.1.2.** Construct a context-free grammar that generates the following strings (and no others):

(a) $0^k 1^k$ (the numbers of zeros and ones are equal);

(b) $0^{2k} 1^k$ (the number of zeros is twice as large as the number of ones);

(c) $0^k 1^l$ (the number of zeros $k$ is larger than the number of ones $l$).

(d) (communicated by M. Sipser) all the strings $X2Y$ where $X$ and $Y$ are composed of 0s and 1s and $X \neq Y$. □

**15.1.3.** Prove that there is no context-free grammar that generates all strings of type $0^k 1^k 2^k$ (and no other strings).

[*Hint.* Prove the following lemma about an arbitrary context-free language: Any sufficiently long string $F$ in the language can be represented as $F = ABCDE$ in such a way that any string $AB^k C D^k E$ (where $B^k$ is $B$ repeated $k$ times) belongs to the language. To prove this lemma, find a nonterminal that is a descendant of itself in the "derivation tree".] □

A nonterminal may be considered a "class name" for all generated strings. In the next example, we use fragments of English words as nonterminals; each fragment is considered to be one nonterminal symbol of the grammar.

*Example 3.*

Terminals:  + * ( ) x

Nonterminals:  ⟨expr⟩ ⟨restexpr⟩ ⟨summ⟩ ⟨restsumm⟩ ⟨fact⟩

Production rules:

$$\langle expr \rangle \rightarrow \langle summ \rangle \langle restexpr \rangle$$

$$\langle restexpr \rangle \rightarrow + \langle expr \rangle$$

$$\langle restexpr \rangle \rightarrow$$

$$\langle summ \rangle \rightarrow \langle fact \rangle \langle restsumm \rangle$$

$$\langle restsumm \rangle \rightarrow * \langle summ \rangle$$

$$\langle restsumm \rangle \rightarrow$$

$$\langle fact \rangle \rightarrow x$$

$$\langle fact \rangle \rightarrow ( \langle expr \rangle )$$

According to this grammar, an **expr**ession is a sequence of **summ**ands separated by symbols +; a **summ**and is a sequence of **fact**ors, separated by symbols *; a **fact**or is either the letter x or an **expr**ession in parentheses.

**15.1.4.** Give another grammar that generates the same language.

*Answer.* Here is one possibility:

$$\langle expr \rangle \rightarrow \langle expr \rangle + \langle expr \rangle$$

$$\langle expr \rangle \rightarrow \langle expr \rangle * \langle expr \rangle$$

$$\langle expr \rangle \rightarrow x$$

$$\langle expr \rangle \rightarrow ( \langle expr \rangle )$$

This grammar is simpler, but not quite as good (see below).                     □

**15.1.5.** An arbitrary context-free grammar is given. Construct an algorithm that checks if an input string belongs to the language generated by the grammar. The algorithm should run in polynomial time: the number of operations should not exceed $P$(input length) for some polynomial $P$. (The polynomial may depend on the grammar.)

*Solution.* The required polynomial time bound rules out any solution based on exhaustive search. However, a polynomial algorithm for a general context-free language exists. We give an outline of that algorithm below. In fact, it has no practical value, because all context-free grammars used in practice have special properties that make more efficient algorithms possible.

(1) Let $K_1, \ldots, K_n$ be the nonterminals of the given grammar. Construct a new context-free grammar with nonterminals $K_1', \ldots, K_n'$. This grammar has the following property: a string $S$ can be generated from $K_i'$ (in the new grammar) if and only if $S$ is nonempty and can be generated from $K_i$ in the old grammar.

To do that, we must know which nonterminals of the given grammar generate the empty string. Then each rule is replaced by a set of rules obtained as follows: On the left-hand side we add the dash, and on the right-hand side we omit some of the nonterminals that generate the empty string and put dashes near the other nonterminals. For example, if the initial grammar has the rule

$$K \to L \; M \; N$$

and the empty string may be generated from L and N but not from M, the new grammar contains rules

$$K' \to L'M'N'$$
$$K' \to M'N'$$
$$K' \to L'M'$$
$$K' \to M'$$

(2) Therefore, we have reduced our problem to the case of a grammar where no terminal generates an empty string. Now we eliminate "cycles" of the form

$$K \to L$$
$$L \to M$$
$$M \to N$$
$$N \to K$$

(each rule has one nonterminal and no terminals on the right-hand side; nonterminals form a cycle of any length). This is easy; we identify all the nonterminals that appear in the same cycle.

(3) Now the membership test for the language generated by a grammar can be performed as follows. For any substring of a given string, we determine which nonterminals can generate this substring. We consider substrings in the order of increasing length. For a given substring, nonterminals are considered in such an order that for any rule of the form $K \to L$, the nonterminal $L$ is considered before the nonterminal $K$. (This is possible because there are no cycles.) Let us explain this process with an example.

Assume that the grammar has rules

$$K \to L$$
$$K \to M \; N \; L$$

and no other rules with K on the left-hand side. We want to know if a given word $A$ may be derived from the nonterminal K. This happens:

- if $A$ can be derived from L;
- if $A$ can be split into $A = BCD$ where $B, C, D$ are nonempty strings such that $B$ can be derived from M, $C$ can be derived from N, and $D$ can be derived from L.

All this information is available because $B, C,$ and $D$ are shorter than $A$ and the nonterminal L is considered before the nonterminal K.

It is easy to see that the running time of the algorithm is polynomial. The degree of the polynomial depends on the number of nonterminals on the right-hand side of the grammar rules. The degree can be made smaller if we convert the grammar into a form where right-hand sides of rules contain not more than two nonterminals. This can be done easily; for example, the rule K → LMK may be replaced by two rules K → LN and N → MK where N is a new nonterminal. □

Some grammars have faster parsing algorithms. We consider several techniques in the next sections (and the next chapter), but it is more instructive to play with some examples now.

**15.1.6.** Consider a grammar with one nonterminal symbol K, terminals 0, 1, 2, and 3, and the rules

$$K \to 0$$
$$K \to 1 \ K$$
$$K \to 2 \ K \ K$$
$$K \to 3 \ K \ K \ K$$

How do we check whether a given string belongs to the corresponding language if the string is scanned from left to right? The number of operations per character should be limited by a constant.

*Solution.* An integer variable n is used along with the invariant relation: *the input string belongs to the language if and only if the non-processed part of the input string is a concatenation of n strings from the language.* □

**15.1.7.** Repeat the previous problem for the grammar

$$K \to 0$$
$$K \to K \ 1$$
$$K \to K \ K \ 2$$
$$K \to K \ K \ K \ 3$$

□

## 15.2 Recursive-descent parsing

Unlike the algorithm of the preceding section (which is of mostly theoretical interest), the recursive-descent parsing algorithm is used quite often. However, it is not applicable to all grammars. (See below the requirements that allow us to apply this method.)

The idea is as follows. For any nonterminal K we construct a procedure ReadK that (being applied to any input string $x$) does two things:

- finds the maximal prefix $z$ of the string $x$ that may appear as a prefix of some string derivable from K;
- says if the string $z$ is derivable from K.

Before we give a more detailed description of this method, we should agree how the procedures access the input string and how they communicate their results. We assume that the input string is read character-by-character. In other words, we assume that there is a separator between the "already read" (processed) part and the "unread" part. (The last name should not be taken literally, because the first symbol of the unread part may be already known to the procedure.)

We assume that there exists a function without parameters

$$\texttt{Next: Symbol}$$

which returns the first symbol of the unread part. Its values are terminals as well as the special symbol EOI that stands for "End Of Input"; this symbol means that the input string is ended. (In a sense, EOI is written after the last character of the input string.) A call to Next does not move the separator between the read and unread parts. There exists a special procedure Move that "reads" the next character; that is, moves the separator to the right, adding one character to the processed part. This procedure is applicable when Next<>EOI. Finally, we have also a Boolean variable b; its role is described below

Now we state our requirements for the procedure ReadK:

- ReadK reads the maximal prefix *A* of the input string (its unprocessed part) that may appear as a prefix of some string derivable from K;
- the value of b becomes true or false depending on whether *A* is derivable from *K* or is only a prefix of some derivable string (but is not derivable itself).

It is convenient to use the following terminology: Any string that is derivable from some nonterminal K is called a K-*string*. Any string that is a prefix of a string derivable from K is called a K-*prefix*. If the two requirements for ReadK stated above are fulfilled, we say that "ReadK is correct for K".

Let us begin with an example. Assume that the rule

$$\texttt{K} \rightarrow \texttt{L M}$$

is the only rule of the grammar that has K on the left-hand side. Assume that L, M are nonterminals and ReadL, ReadM are correct procedures for those nonterminals.

Consider the following procedure:

```
procedure ReadK;
begin
  ReadL;
  if b then begin
    ReadM;
  end;
end;
```

**15.2.1.** Give an example where this procedure is not correct for K.

*Answer.* Assume that any string 000...000 is derivable from L and that only the string 01 is derivable from M. Then the string 00001 is derivable from K, but the procedure ReadK does not see this.                                                    □

Let us give a sufficient condition for the correctness of the procedure ReadK given above. To do that, we need some notation. Assume that a context-free grammar is fixed and that $N$ is some nonterminal of that grammar. Consider the $N$-string $A$ that has a proper prefix $B$, which is also an $N$-string (assuming such $A$ and $B$ exist). For each pair of such $A$ and $B$, consider the terminal that follows $B$ in $A$ (appears immediately after $B$ in $A$). The set of all such symbols (for all $A$ and $B$) is denoted by Foll($N$). (If no $N$-string is a proper prefix of another $N$-string, the set Foll($N$) is empty.)

**15.2.2.** Find (a) Foll(E) for the grammar given in example 1 (p. 222); (b) Foll(E) and Foll(T) for the grammar give in example 2 (p. 222); (c) Foll($\langle$summ$\rangle$) and Foll($\langle$fact$\rangle$) for the grammar given in example 3 (p. 223).

*Answer.* (a) Foll(E) = { [, (}. (b) Foll(E) = { [, (}; Foll(T) is empty (no T-string is a prefix of another T-string). (c) Foll($\langle$summ$\rangle$) = {*}; Foll($\langle$fact$\rangle$) is empty.      □

For any nonterminal $N$, we denote the set of all terminals that are first characters of nonempty $N$-strings by First($N$). Now we are ready to give a sufficient condition for the correctness of the procedure ReadK in the situation explained above.

**15.2.3.** Prove that if Foll(L) and First(M) are disjoint and the set of all M-words is not empty, then the procedure ReadK is correct for K.

*Solution.* Consider two cases.

(1) Suppose that after the call to ReadL the value of b is false. In this case, ReadL reads the maximal L-prefix $A$; this prefix is not an L-string. The string $A$ is a K-prefix (here we use the fact that the set of strings derivable from M is not empty). Will $A$ be the maximal prefix of the input string that is at the same time a K-prefix? The answer is "yes". Indeed, assume that $A$ is not maximal and there exists a longer string $X$ that is both a K-prefix and a prefix of the input string. Since ReadL is correct, $X$ is not a K-prefix, and therefore, $X = BC$ where $B$ is an L-string and $C$ is a M-prefix.

If $B$ is longer than $A$, then $A$ is not the maximal prefix of the input string that is also a K-prefix, which contradicts the correctness of ReadL. If $B = A$, then $A$ would be an L-string, which is not true. Therefore, $B$ is a proper prefix of $A$, $C$ is not empty, and the first character of $C$ follows the last character of $B$ in $A$. So the first character of $C$ belongs both to Foll(L) and First(M), which contradicts our assumption.

This contradiction shows that $A$ is a maximal prefix of the input string that is also a K-prefix. Moreover, the argument above shows that $A$ is not a K-string. The correctness of the procedure ReadK is therefore established.

(2) Assume that after the call to ReadL, the value of b is true. Then the procedure ReadK reads some string of the form $AB$ where $A$ is an L-string and $B$ is an M-prefix. Therefore, $AB$ is a K-prefix. Let us check that it is maximal. Assume that $C$ is a longer prefix, which is at the same time a K-prefix. Then either $C$ is an L-prefix (which is impossible because $A$ is the maximal L-prefix) or $C = A'B'$, where $A'$ is an

L-string and $B'$ is an M-prefix. If $A'$ is shorter than $A$, then $B'$ is not empty and begins with a character that belongs both to First(M) and Foll(L), which is impossible. If $A'$ is longer than $A$, then $A$ is not the maximal L-prefix. Therefore, the only possibility is $A' = A$, but in this case $B$ is a prefix of $B'$, which contradicts the correctness of ReadM. Therefore, $AB$ is the maximal prefix of the input string that is a K-prefix.

It remains to check that the value of b returned by ReadK is correct. If b is true, this is evident. If b is false, then $B$ is not an M-string, and we have to check that $AB$ is not a K-string. Indeed, if $AB = A'B'$ where $A'$ is an L-string and $B'$ is an M-string, then $A'$ cannot be longer than $A$ (since ReadL reads the maximal prefix), $A'$ cannot be equal to $A$ (since in this case $B'$ would be equal to $B$ and could not be an M-string), and $A'$ cannot be shorter than $A$ (since in this case the first character of $B'$ would belong both to First(M) and Foll(L)). The correctness of ReadK is proved. $\square$

Now we consider another special case. Assume that a context-free grammar contains the rules

$$K \to L$$
$$K \to M$$
$$K \to N$$

and has no other rules with K on the left-hand side.

**15.2.4.** Assume that ReadL, ReadM, and ReadN are correct (for L, M, and N) and that First(L), First(M), and First(N) are disjoint. Write a procedure ReadK that is correct for K.

*Solution.* Here is the procedure:

```
procedure ReadK;
begin
  if (Next is in First(L)) then begin
    ReadL;
  end else if (Next is in First(M)) then begin
    ReadM;
  end else if (Next is in First(N)) then begin
    ReadN;
  end else begin
    b := true or false depending on whether an
        empty string is derivable from K or not
  end;
end;
```

Let us prove that ReadK is correct for K. If the symbol Next is not in the sets First(L), First(M), and First(N), then the empty string is the maximal prefix of the input string that is a K-prefix. If Next belongs to one of those sets (and, therefore, does not belong to the others), then the maximal prefix of the input string that is a K-prefix is nonempty and the corresponding procedure reads it. $\square$

**15.2.5.** Using the methods discussed, write a procedure that recognizes expressions generated by the grammar of example 3 (p. 223):

$$\langle expr \rangle \rightarrow \langle summ \rangle \, \langle restexpr \rangle$$

$$\langle restexpr \rangle \rightarrow + \langle expr \rangle$$

$$\langle restexpr \rangle \rightarrow$$

$$\langle summ \rangle \rightarrow \langle fact \rangle \, \langle restsumm \rangle$$

$$\langle restsumm \rangle \rightarrow * \langle summ \rangle$$

$$\langle restsumm \rangle \rightarrow$$

$$\langle fact \rangle \rightarrow x$$

$$\langle fact \rangle \rightarrow ( \, \langle expr \rangle \, )$$

*Solution.* This grammar does not follow the patterns above: among the right-hand sides of its rules there are combinations of terminals and nonterminals such as

$$+ \, \langle expr \rangle$$

as well as a group of three symbols

$$( \, \langle expr \rangle \, )$$

This grammar also contains several rules with the same left-hand side and right-hand sides of different types, such as

$$\langle restexpr \rangle \rightarrow + \langle expr \rangle$$

$$\langle restexpr \rangle \rightarrow$$

These problems are not fatal. For example, a rule of type $K \rightarrow L \, M \, N$ may be replaced by two rules $K \rightarrow L \, Q$ and $Q \rightarrow M \, N$. The terminals on the right-hand side may be replaced by nonterminals (the only rule involving these nonterminals allows us to replace them by the corresponding terminals). If several rules have the same left-hand side and different right-hand sides, such as

$$K \rightarrow L \, M \, N$$

$$K \rightarrow P \, Q$$

$$K \rightarrow$$

they can be replaced by rules

$$K \rightarrow K_1$$

$$K \rightarrow K_2$$

$$K \rightarrow K_3$$

$$K_1 \rightarrow L \, M \, N$$

$$K_2 \rightarrow P \, Q$$

$$K_3 \rightarrow$$

We will not, however, transform the grammar of example 3 explicitly. Instead, we imagine that this transformation is performed (new nonterminals added), then the procedures for all nonterminals (old and new) are written, and finally the procedures for the new nonterminals are eliminated (by in-line substitutions). For example, for the rule

$$K \rightarrow L \ M \ N$$

we get the procedure

```
procedure ReadK;
begin
  ReadL;
  if b then begin ReadM; end;
  if b then begin ReadN; end;
end;
```

Its correctness is guaranteed if (1) Foll(L) and First(MN) are disjoint (First(MN) is equal to First(M) if the empty string is not derivable from M; otherwise, it is equal to the union of First(M) and First(N)); (2) Foll (M) and First(N) are disjoint.

Similarly, the rules

$$K \rightarrow L \ M \ N$$
$$K \rightarrow P \ Q$$
$$K \rightarrow$$

lead to the procedure

```
procedure ReadK;
begin
  if (Next is in First(LMN)) then begin
    ReadL;
    if b then begin ReadM; end;
    if b then begin ReadN; end;
  end else if (Next is in First(PQ)) then begin
    ReadP;
    if b then begin ReadQ; end;
  end else begin
    b := true;
  end;
end;
```

To prove its correctness, we require the sets First(LMN) and First(PQ) to be disjoint.

Now we apply these methods to the grammar of example 3:

```
procedure ReadSymb (c: Symbol);
  b := (Next = c);
  if b then begin Move; end;
end;
```

```
procedure ReadExpr;
  ReadSumm;
  if b then begin ReadRestExpr; end;
end;

procedure ReadRestExpr;
  if Next = '+' then begin
    ReadSymb ('+');
    if b then begin ReadExpr; end;
  end else begin
    b := true;
  end;
end;

procedure ReadSumm;
  ReadFact;
  if b then begin ReadRestSumm; end;
end;

procedure ReadRestSumm;
  if Next = '*' then begin
    ReadSymb ('*');
    if b then begin ReadSumm; end;
  end else begin
    b := true;
  end;
end;

procedure ReadFact;
  if Next = 'x' then begin
    ReadSymb ('x');
  end else if Next = '(' then begin
    ReadSymb ('(');
    if b then begin ReadExpr; end;
    if b then begin ReadSymb (')'); end;
  end else begin
    b := false;
  end;
end;
```

These procedures are mutually recursive; that is, some procedure uses another one which in its turn uses the first one, etc. This is allowed in Pascal if we use the so-called forward definitions of the mutually recursive procedures. As usual, to prove the correctness of recursive procedures we need to prove that (1) each of them is correct, assuming all calls work correctly (here our method works: one needs only to check that the corresponding sets are disjoint); (2) the procedure terminates. The second claim is not self-evident. For example, if the grammar has the rule K → KK, then no strings are derivable from K, and the sets Foll (K) and First (K) are empty (and therefore disjoint), but the procedure

```
procedure ReadK;
begin
  ReadK;
  if b then begin
    ReadK;
  end;
end;
```

(written according to our guidelines) never terminates.

In the case in question, the procedures `ReadRestExpr`, `ReadRestAdd`, and `ReadFact` either terminate immediately or decrease the length of the unprocessed part of the input string. Since any cycle of the mutually recursive calls includes one of them, termination is guaranteed. Our problem is solved. □

**15.2.6.** Assume that a grammar has two rules with nonterminal K on the left-hand side:

$$K \to L \ K$$

$$K \to$$

According to these rules, any K-string is a concatenation of several L-strings. Assume also that the sets Foll(L) and First(K) (which equals First(L) in this case) are disjoint. Assume that a procedure `ReadL` is correct for L. Write a *non-recursive* procedure `ReadK` that is correct for K.

*Solution.* As we already know, the following recursive procedure is correct for K:

```
procedure ReadK;
begin
  if (Next is in First(L)) then begin
    ReadL;
    if b then begin ReadK; end;
  end else begin
    b := true;
  end;
end;
```

Termination is guaranteed because the length of the unprocessed part is decreased before the recursive call.

This recursive procedure is equivalent to the following non-recursive one:

```
procedure ReadK;
begin
```

```
  b := true;
  while b and (Next is in First(L)) do begin
  | ReadL;
  end;
end;
```

Let us formally check this equivalence. Termination is guaranteed both for the recursive and non-recursive procedures. Therefore, it is enough to check that the body of the recursive procedure becomes equivalent to the body of the non-recursive one if the recursive call is replaced by the call of the non-recursive procedure. Let us make this replacement:

```
if (Next is in First(L)) then begin
  ReadL;
  if b then begin
    b := true;
    while b and (Next is in First(L)) do begin
    | ReadL;
    end;
  end;
end else begin
| b := true;
end;
```

The first command b:=true may be deleted because at this point b is already true. The second command b:=true may be moved to the beginning:

```
b := true;
if (Next is in First(L) then begin
  ReadL;
  if b then begin
    while b and (Next is in First(L)) do begin
    | ReadL;
    end;
  end;
end;
```

Now the second if may be removed (because if b is false, the while-loop does nothing). We may also add the condition b to the first if (because b is true at that point). Thus we get

```
b := true;
if b and (Next is in First(L)) then begin
  ReadL;
  while b and (Next is in First(L)) do begin
  | ReadL;
  end;
end;
```

which is equivalent to the body of the non-recursive procedure above (the first iteration of the loop is unfolded). ☐

**15.2.7.** Prove the correctness of the non-recursive procedure shown above directly, without referring to the recursive version.

*Solution.* Consider the maximal prefix of the input string that is a K-prefix. It can be represented as a concatenation of several nonempty strings: all are L-strings except, maybe, the last one, which is an L-prefix. We call those strings (including the last one) "components".

The invariant relation: *several components have been read; b is true if and only if the last component is an L-string.*

Let us check that this invariant relation remains true after the next iteration. If only the last component remains, it is evident. If several components remain, the first of the remaining components is followed by a character that belongs to First(L) and is therefore not in Foll(L); so the first remaining component is a maximal L-prefix that is also a prefix of the unprocessed part. ☐

In practice a shorthand notation for grammars is used. Namely, rules of the form

$$K \to L \ K$$

$$K \to$$

(we assume that no other rule has K on the left-hand side, so K-strings are concatenations of L-strings) are omitted, and K is replaced by L enclosed in curly braces (which denotes iteration in this case). Also, several rules with the same left-hand side are often written as one rule where alternatives are written one after another separated by bars.

For example, the grammar for **expr**essions given above may be rewritten as follows:

$$\langle \text{expr} \rangle \to \langle \text{summ} \rangle \ \{ \ + \ \langle \text{summ} \rangle \ \}$$

$$\langle \text{summ} \rangle \to \langle \text{fact} \rangle \ \{ \ * \ \langle \text{fact} \rangle \ \}$$

$$\langle \text{fact} \rangle \to x \ | \ ( \ \langle \text{expr} \rangle \ )$$

**15.2.8.** Write a procedure that is correct for ⟨expr⟩, following this grammar. Use iteration instead of recursion whenever possible.

*Solution.*

```
procedure ReadSymb (c: Symbol);
  b := (Next = c);
  if b then begin Move; end;
end;

procedure ReadExpr;
begin
  ReadSumm;
```

```
    while b and (Next = '+') do begin
    | Move; ReadSumm;
    | end;
end;

procedure ReadSumm;
begin
| ReadFact;
| while b and (Next = '*') do begin
| | Move; ReadFact;
| | end;
end;

procedure ReadFact;
begin
| if Next = 'x' do begin
| | Move; b := true;
| end else if Next = '(' then begin
| | Move; ReadExpr;
| | if b then begin ReadSymb (')'); end;
| end else begin
| | b := false;
| | end;
end;                                                          □
```

**15.2.9.** The assignment b:=true in the last procedure may be omitted. Why?

*Solution.* We may assume that all procedures are called only when b=true.    □


## 15.3 Parsing algorithm for LL(1)-grammars

In this section, we consider one more algorithm to check if a given string can be generated by a given grammar. This algorithm is called LL(1)-*parsing*. Its main idea can be summed up in one sentence: we may assume that all the production rules are applied to the leftmost nonterminal only; if we are lucky, the applicable rule is determined uniquely by the first character of the string derivable from this nonterminal.

Now we give the details. To begin with, we have the following:

*Definition.* A *leftmost derivation* (of a string in a grammar) is a derivation where the leftmost nonterminal is replaced at each step.

**15.3.1.** Each derivable word (which contains only terminals) has the leftmost derivation.

*Solution.* During the derivation process, different nonterminals in a string are replaced independently. (That is why the grammar is called "context-free".) In other words, if at some point of the derivation we have the string $\ldots K \ldots L \ldots$ where $K$ and $L$ are nonterminals, then the substitutions for $K$ and $L$ may be performed in any order. Therefore, we can rearrange the derivation in such a way that the left nonterminal $K$ is replaced first. □

**15.3.2.** Consider the grammar with four production rules:

$$(1) \quad E \rightarrow$$
$$(2) \quad E \rightarrow TE$$
$$(3) \quad T \rightarrow (E)$$
$$(4) \quad T \rightarrow [E]$$

Find the leftmost derivation of the word $A = [()([])]$ and prove that it is unique.

*Solution.* At the first step, only rule (2) may be applied:

$$E \rightarrow TE$$

What happens with T then? Since $A$ starts with [, only rule (4) can be applied:

$$E \rightarrow TE \rightarrow [E]E$$

The leftmost E is now replaced by TE (otherwise the second symbol of the input string would be ]):
$$E \rightarrow TE \rightarrow [E]E \rightarrow [TE]E$$

and T is replaced according to (3):

$$E \rightarrow TE \rightarrow [E]E \rightarrow [TE]E \rightarrow [(E)E]E$$

Now the leftmost E should be replaced by the empty string, otherwise the third character of the input string would be ( or [ (other characters cannot be the first character of a T-string):

$$E \rightarrow TE \rightarrow [E]E \rightarrow [TE]E \rightarrow [(E)E]E \rightarrow [()E]E$$

We continue:

$$\ldots \rightarrow [()TE]E \rightarrow [()(E)E]E \rightarrow [()(TE)E]E \rightarrow [()([E]E)E]E \rightarrow$$

$$\rightarrow [()([]E)E]E \rightarrow [()([])E]E \rightarrow [()([])]E \rightarrow [()([])]$$

Thus we see that the leftmost derivation is unique. □

What are the requirements for a grammar that make this approach (finding the unique leftmost derivation) possible? Assume that at some point the leftmost nonterminal is $K$. In other words, we have the string $AKU$ where $A$ is a string containing only terminals and $U$ is a string that may contain both terminals and nonterminals. Suppose the grammar has the rules

$$K \rightarrow L M N$$
$$K \rightarrow P Q$$
$$K \rightarrow R$$

and we have to choose one of them. We make the choice based on the first symbol of the part of the input string that is derivable from $KU$.

Consider the set First($LMN$) of all terminals that are first symbols of nonempty strings of terminals derivable from $LMN$. This set is equal to the union of the set First($L$), the set First($M$) (if the empty string is derivable from $L$), and the set First($N$) (if the empty string is derivable from both $L$ and $N$). To make the choice (based on the first character) possible, we require that the sets First($LMN$), First($PQ$), and First($R$) are disjoint. But this is not the only requirement. Indeed, it is possible, for example, that the empty string is derivable from $LMN$, and the string derived from $U$ starts with a character in First($PQ$). The definitions below take this problem into account.

A language recognized by a context-free grammar was defined as the set of all strings *of terminals* derivable from the initial nonterminal (axiom). We will also speak about strings composed of terminals and nonterminals derivable from the axiom, or from any other nonterminal, or from any string composed of terminals and nonterminals. So the relation "derivable from" can be considered as a binary relation defined on the set of all strings composed of terminals and nonterminals. (However, if we say that some string is derivable and do not specify the starting point of the derivation, we always mean that the derivation starts from the axiom.)

For any string $X$ composed of terminal and nonterminals, First($X$) denotes the set of all terminals that are the first characters of nonempty strings of terminals derivable from $X$. If for any nonterminal there is at least one string of terminals derivable from it, then the phrase "of terminals" may be omitted in the definition. We assume in the sequel that this condition is satisfied.

For any nonterminal $K$, the notation Follow($K$) is used for the set of all terminals that appear in the derivable (from the axiom) strings immediately after $K$. (Please do not confuse this set with Foll($K$) defined in the preceding section.) We add the symbol EOI to Follow($K$) if there exists a derivable string that ends with $K$.

For each rule

$$K \rightarrow V$$

(where $K$ is a nonterminal and $V$ is a string that contains terminals and nonterminals) we define the set of *leading terminals*, which is denoted by Lead($K \rightarrow V$). By definition, Lead($K \rightarrow V$) is equal to First($V$) or the union of First($V$) and Follow($K$) if the empty string is derivable from $V$.

*Definition.* A context-free grammar is called an LL(1)-*grammar* if for any two rules $K \rightarrow V$ and $K \rightarrow W$ with the same left-hand sides, the sets Lead($K \rightarrow V$) and Lead($K \rightarrow W$) are disjoint.

**15.3.3.** Is the grammar

$$K \rightarrow K \ \#$$

$$K \rightarrow$$

(derivable strings are sequences of #'s) an LL(1)-grammar?

*Solution.* No, because # is a leading terminal for both rules. (This is true for the second rule because # belongs to Follow(K).) □

**15.3.4.** Write an equivalent LL(1)-grammar.

*Solution.*

$$K \rightarrow \# \ K$$

$$K \rightarrow$$

We have replaced a "left-recursive" rule by a "right-recursive" one. □

The next problem shows that for an LL(1)-grammar, the next step in the construction of a leftmost derivation is uniquely defined.

**15.3.5.** Assume that a string $X$ is derivable in an LL(1)-grammar and $K$ is the leftmost nonterminal in $X$; that is, $X = AKS$ where $A$ is a string of terminals and $S$ is a string of terminals and nonterminals. Assume that two different rules of the grammar have $K$ on the left-hand side, and both of them were applied to the nonterminal $K$ selected in $X$. Both derivations were continued and two strings of terminals (having prefix $A$) were obtained. Prove that this prefix is followed by different terminals. (Here we consider EOI as a terminal.)

*Solution.* Those terminals are leading terminals of two different rules. □

**15.3.6.** Prove that if a string is derivable in an LL(1)-grammar, its leftmost derivation is unique.

*Solution.* The preceding problem shows that at each step there is only one possible continuation. □

**15.3.7.** A grammar is called *left-recursive* grammar if there exists a nonterminal $K$ and a string derivable from $K$ that starts with $K$ (but is not equal to $K$). Prove that if (1) a grammar $G$ is left-recursive; (2) for each nonterminal $K$, there exists a nonempty string derivable from $K$; and (3) for each nonterminal $K$, there exists a derivation starting from the axiom and including $K$, then $G$ is not an LL(1)-grammar.

*Solution.* Consider the derivation of a string $KU$ from a nonterminal $K$ where $U$ is a nonempty string. We may assume that it is a leftmost derivation (other nonterminals may remain untouched). Consider the derivation

$$K \rightsquigarrow KU \rightsquigarrow KUU \rightsquigarrow \cdots$$

(here $\rightsquigarrow$ stands for several derivation steps) and the derivation $K \rightsquigarrow A$ where $A$ is a nonempty string of terminals. At some point these two derivations diverge; however, both derivations may lead to a string that starts with $A$ (in the first derivation there is still the nonterminal $K$ at the beginning, which may be transformed to $A$). This contradicts the fact that the next step of the leftmost derivation is determined uniquely by the first character of the derived string. (This uniqueness is valid for derivations that start from the axiom; recall that the nonterminal $K$ may appear in such a derivation by assumption.)                                                                                                $\square$

Therefore, the LL(1) approach cannot be applied to left-recursive grammars (except for trivial cases). We have to transform them to equivalent LL(1)-grammars first (or use other parsing algorithms).

**15.3.8.** For any LL(1)-grammar, construct an algorithm that checks if the input string belongs to the language generated by the grammar. Use the preceding results.

*Solution*. We follow the scheme outlined above and look for a leftmost derivation of the given string. At each point, we have an initial part of the leftmost derivation that ends with a string composed of terminals and nonterminals. This string has the processed part of the input string as a prefix. Our algorithm stores the remaining part. In other words, we keep a string S of terminals and nonterminals with the following properties (the processed part of the input string is denoted by A):

1. the string AS is derivable (in the grammar);
2. any leftmost derivation of the input string includes the string AS.

These properties are denoted by "(I)" in the sequel.

Initially, A is empty and S contains only one nonterminal (the axiom).

If at some point the string S begins with a terminal t and t = Next, then we may call the procedure Move and delete the initial terminal t from S. Indeed, this operation leaves AS unchanged.

If the string S starts with a terminal t and t ≠ Next, then the input string is not derivable at all, because (I) implies that any (leftmost) derivation goes through the stage AS. (The same is true if Next = EOI.)

If S is empty, the condition (I) implies that the input string is derivable if and only if Next = EOI.

The only remaining case is that S starts with some nonterminal K. As we have already shown, all the leftmost derivations that start with S and end with a string whose first character is Next, begin with the same production rule; that is, the production rule whose set of leading terminals includes Next. If such a rule does not exist, the input string is not derivable at all. If such a rule exists, we apply it to the opening nonterminal K of the string S and property (I) remains valid. We arrive at the following algorithm:

```
S := empty string;
error := false;
{error => input string is not derivable}
{not error => (I)}
while (not error) and not ((Next=EOI) and (S is empty))
      do begin
    if (S starts with a terminal equal to Next) then begin
```

```
  |   Move; delete the first symbol from S;
  end else if (S starts with a terminal <> Next)
  |     then begin
  |   error := true;
  end else if (S is empty) and (Next <> EOI) then begin
  |   error := true;
  end else if (S starts with some nonterminal K and Next
  |       belongs to the set of leading terminals for one of
  |       the production rules for K) then begin
  |   apply this rule to K
  end else if (S starts with some nonterminal K and Next
  |       does not belong to the set of leading terminals
  |       for all the production rules for K) then begin
  |   error := true;
  end else begin
  |   {this cannot happen}
  end;
end;
{the input string is derivable <=> not error}
```

This algorithm always terminates. Indeed, if a terminal appears as the first symbol in S, the algorithm stops or reads the next input character. If nonterminals alternate as first symbols of S in an infinite loop, then the grammar is left-recursive; we may assume that this is not the case. (This follows from the preceding problem; we may easily remove from the grammar all the nonterminals that do not appear in derivations beginning with the axiom; the same can be done for nonterminals from which only the empty string is derivable.) □

*Remarks.*

- This algorithm uses S as a stack (all operations are made near its left end).
- In either of the last two cases (in the if-construct), no input characters are read. Therefore, we can precompute the action for all nonterminals and all possible values of Next. Doing that, we need only one iteration per input character.
- In practice, it is convenient to have a table that lists all actions for all pairs (input symbol, nonterminal), and a small program that interprets this table.

**15.3.9.** To check if a given grammar is an LL(1)-grammar, we need to compute Follow($T$) and First($T$) for all nonterminals $T$. How can we do that?

*Solution.* If the grammar includes, say, the rule $K \to L\,M\,N$, then ($\Lambda$ denotes the empty string):

$$\text{First}\,(L) \subset \text{First}\,(K),$$
$$\text{First}\,(M) \subset \text{First}\,(K), \quad \text{if } \Lambda \text{ is derivable from } L,$$
$$\text{First}\,(N) \subset \text{First}\,(K), \quad \text{if } \Lambda \text{ is derivable both from } L \text{ and } M,$$
$$\text{Follow}\,(K) \subset \text{Follow}\,(N),$$
$$\text{Follow}\,(K) \subset \text{Follow}\,(M), \text{ if } \Lambda \text{ is derivable from } N,$$
$$\text{Follow}\,(K) \subset \text{Follow}\,(L), \quad \text{if } \Lambda \text{ is derivable both from } M \text{ and } N,$$
$$\text{First}\,(N) \subset \text{Follow}\,(M),$$
$$\text{First}\,(M) \subset \text{Follow}\,(L),$$
$$\text{First}\,(N) \subset \text{Follow}\,(L), \quad \text{if } \Lambda \text{ is derivable from } M.$$

These rules (written for all productions) allow us to generate the sets $\text{First}(T)$, and thereafter $\text{Follow}(T)$, for all terminals and nonterminals $T$. As a starting point we use

$$\text{EOI} \in \text{Follow}\,(K)$$

for an initial nonterminal $K$ (the axiom) and

$$z \in \text{First}\,(z)$$

for any terminal $z$. We stop the generation process when the repeated applications of the rules give no new elements of the sets $\text{First}(T)$ and $\text{Follow}(T)$. □

# 16

# Left-to-right parsing (LR)

In this chapter we consider another approach to parsing, called an LR(1)-parsing algorithm, as well as some simplified versions of it. We start by describing a general scheme of left-to-right parsing (section 16.1). Then we consider a class of grammars for which this scheme can be implemented easily (LR(0)-grammars, section 16.2). Some extensions of this class are discussed in sections 16.3 (SLR(1)-grammars) and 16.4 (LR(1)- and LALR(1)-grammars). We conclude the chapter with some general remarks about parsing (section 16.5).

## 16.1 LR-processes

There are two main differences between LR(1)-parsing and LL(1)-parsing. First, we seek a *rightmost* derivation, not a leftmost one. Second, we construct the derivation from the bottom (beginning with the input string) to the top (the axiom) and not vice-versa (as in LL(1)-parsing).

A *rightmost* derivation is a derivation where the rightmost nonterminal is replaced at each step.

**16.1.1.** Prove that any derivable string of terminals has a rightmost derivation.□

It is convenient to look at the rightmost derivation backwards, starting from the input string. Let us define the notion of an LR-*process* on the input string $A$. This process involves the string $A$ and another string $S$ that contains both terminals and nonterminals. Initially, the string $S$ is empty. The LR-process includes two types of actions:

(1) the first character of $A$ (called the next input symbol and denoted by Next) may be moved to the end of the string $S$ (and deleted from $A$); this action is called a *shift* action;

(2) if the right-hand side of some production rule is a suffix of $S$, then it can be replaced by the nonterminal that is on the left-hand side of that rule; the string $A$ remains unchanged. This action is called a *reduce* action.

A. Shen, *Algorithms and Programming*, Springer Undergraduate Texts in Mathematics and Technology, DOI 10.1007/978-1-4419-1748-5_16, © Springer Science+Business Media, LLC 2010

Let us mention that the LR-process is not deterministic; there are situations where many different actions are possible.

We say that the LR-process on a string $A$ is *successful* if the string $A$ becomes empty and the string $S$ contains only one nonterminal, and this nonterminal is the initial nonterminal (the axiom).

**16.1.2.** Prove that for any string $A$ (of terminals) a successful LR-process exists if and only if $A$ is derivable in the grammar. Find a one-to-one correspondence between rightmost derivations and successful LR-processes.

*Solution.* The shift action does not change the string $SA$. The reduce action changes $SA$ and this change is a reversed step of a derivation. This derivation is a rightmost one because the reduction is done at the end of $S$ and all symbols of $A$ are terminals. Therefore, each LR-process corresponds to a rightmost derivation.

Conversely, assume that a rightmost derivation is given. Imagine a separator placed after the last nonterminal in the string. When a production rule is applied to that nonterminal, we may need to move the separator to the left (if the right-hand side of the rule applied ends with a terminal). Splitting this move into steps (one symbol per step) we get a process that is exactly an inverted LR-process.  □

All changes in the string $S$ during an LR-process are made near its right end. This is why the string $S$ is called the *stack* of the LR-process.

So the problem of finding the rightmost derivation of a given string is the problem of constructing a successful LR-process on this string. At each step we have to decide whether we want to apply a shift or reduce action, and choose a production rule if several reductions are possible. In the LR(1)-algorithm, the decision is made based on $S$ and the first symbol of $A$. If only information about $S$ is used, it is an LR(0)-algorithm. (The exact definitions are given below.)

Assume that a grammar is fixed. In the sequel, we assume that for each nonterminal there exists a string of terminals derivable from it.

Let $K \rightarrow U$ be one of the grammar's rules ($K$ is a nonterminal, $U$ is a string of terminals and nonterminals). We consider a set of strings (composed of both terminals and nonterminals) called the *left context* of the rule $K \rightarrow U$. (Notation: LeftCont($K \rightarrow U$).) By definition, this set contains all the strings that may appear as a stack content immediately before the reduction of $U$ to $K$ in a successful LR-process.

**16.1.3.** Reformulate this definition in terms of rightmost derivations.

*Solution.* Consider all rightmost derivations of the form

$$\langle \text{axiom} \rangle \rightsquigarrow \text{XKA} \rightarrow \text{XUA},$$

where $A$ is a string of terminals, $X$ is a string of terminals and nonterminals, and $K \rightarrow U$ is a production rule. All strings $XU$ that appear in those derivations form the left context of the rule $K \rightarrow U$. Indeed, recall that we assume that for any nonterminal there exists a string of terminals derivable from it; therefore, the rightmost derivation of the string $XUA$ may be continued until a right derivation of some string of terminals is obtained.  □

**16.1.4.** All strings from LeftCont(K→U) end with U. Prove that if we delete this suffix U, the resulting set of strings does not depend on which rule (for the nonterminal K) is chosen. This set is denoted by Left(K).

*Solution.* The preceding problem shows that Left(K) is the set of all strings that may appear at the left of the rightmost nonterminal K in some rightmost derivation. □

**16.1.5.** Prove that in the last sentence the words "the rightmost nonterminal" may be omitted: Left(K) is the set of all strings that may appear on the left of any occurrence of K in a rightmost derivation.

*Solution.* The derivation may be continued and all nonterminals on the right of K may be replaced by terminals; this replacement does not change anything on the left of K. □

**16.1.6.** Let $G$ be a grammar. Construct a new grammar $G^l$ such that for any nonterminal K of $G$, the grammar $G^l$ contains a nonterminal ⟨LeftK⟩, and all elements of Left(K) (and no others) are derivable from ⟨LeftK⟩ in $G^l$. The terminals of $G^l$ are nonterminals and terminals of $G$.

*Solution.* Let P be the initial nonterminal of $G$. The new grammar $G^l$ has a rule

$$⟨\text{LeftP}⟩ → \qquad \text{(right-hand side is the empty string)}$$

For any production rule of the $G$, say,

$$\text{K} → \text{L t M N} \qquad \text{(L, M, N are nonterminals, t is a terminal)}$$

we add the following rules to $G^l$:

$$⟨\text{LeftL}⟩ → ⟨\text{LeftK}⟩$$
$$⟨\text{LeftM}⟩ → ⟨\text{LeftK}⟩ \text{L t}$$
$$⟨\text{LeftN}⟩ → ⟨\text{LeftK}⟩ \text{L t M}$$

The meaning of the new rules may be explained as follows. An empty string may appear on the left of P. If a string X may appear on the left of K, then X may appear on the left of L; at the same time XLt may appear on the left of M, and XLtM may appear on the left of N. By induction over the length of a rightmost derivation, we check that everything that may appear on the left of some nonterminal, appears according to these rules. □

**16.1.7.** Why is it important in the preceding problem that we consider only the rightmost derivations?

*Solution.* Otherwise we must take into account transformations performed on the left of K. □

**16.1.8.** A context-free grammar is given. Construct an algorithm that for any input string finds all the sets Left(K) containing the string.

*Remark* (for experts only). The existence of such an algorithm, even a finite automaton (an inductive extension with a finite number of values, see section 1.3), follows from the preceding problem. Indeed, the grammar constructed has a special form: The right-hand sides of rules contain only one nonterminal and it is in the leftmost position. Nevertheless, we give an explicit construction of that automaton below.

*Solution.* By a *situation* of a given grammar we mean one of its rules with some additional information; namely, one of the positions on the right-hand side (before the first symbol, between the first and the second symbols, ..., after the last symbol) is marked. For example, the rule

$$K \to L t M N$$

(K, L, M, N are nonterminals, t is a terminal) gives five situations

$$K \to \_L t M N \quad K \to L \_t M N \quad K \to L t \_M N \quad K \to L t M \_N \quad K \to L t M N \_$$

(the position is indicated by the underscore sign).

We say that a string S is *coherent* with a situation $K \to U \_V$ if S ends with U; that is, if S = TU for some T and, moreover, T belongs to Left(K). (The meaning of this definition may be explained as follows: the suffix U of the stack S is ready for the future reduction of UV into K.) Now we can give an equivalent definition of LeftCont(K → X) as the set of all strings that are coherent with the situation $K \to X \_$, and Left(K) as the set of all strings coherent with the situation $K \to \_X$ (here K → X is any production rule for nonterminal K).

Here is an equivalent definition in terms of LR-processes: S is coherent with the situation $K \to U \_V$ if there exists a successful LR-process such that:

- during the process, the string S appears in the stack and S ends with U;
- for some time S is not touched and the string V appears on the right of S;
- UV is reduced into K;
- the LR-process continues and eventually terminates successfully.

**16.1.9.** Prove the equivalence of these two definitions.

[*Hint.* If S = TU and T belongs to Left(K), then it is possible to have T on the stack, then add U, then V, then reduce UV to K, and finally finish the LR-process successfully. (Several times we use the assumption that for any nonterminal there exists some string of terminals derivable from it; this assumption guarantees that we may add an arbitrary string to the stack.)]                                    □

Our goal is to construct an algorithm that finds all K such that the input string belongs to Left(K). Consider a function that maps each string S (of terminals and nonterminals) into the set of all situations that are coherent with S. This set is called a *state corresponding to* S. We denote it by State(S). It is enough to show that the function State(S) is inductive; that is, the value State(SJ) for any terminal or nonterminal J is determined by State(S) and J. (We have seen that membership in Left(K) may be expressed in terms of that function.) Indeed, the value State(SJ) can be computed according to the following rules (1)–(3):

(1) If the string S is coherent with the situation K → U_V, and the string V starts with the symbol J; that is, V = JW, then SJ is coherent with the situation K → UJ_W.

This rule determines completely what situations not starting with an underscore are coherent with SJ. It remains to find for which nonterminals K the string SJ belongs to Left(K). This can be done according to the following rules:

(2) If the situation L → U_V turns out to be coherent with SJ (according to (1)) and V starts with a nonterminal K, then SJ belongs to Left(K).

(3) If SJ is in Left(L) for some L, the grammar contains a production rule L → V and V starts with a nonterminal K, then SJ belongs to Left(K).

Please note that the rule (3) may be considered a version of rule (2). Indeed, if the assumptions of (3) are valid, then the situation L → _V is coherent with SJ and V starts with a nonterminal K.

The correctness of these rules becomes more or less evident upon reflection. The only thing that requires comment is why rules (2) and (3) generate *all* terminals K such that SJ belongs to Left(K). Let us try to explain why. Consider a rightmost derivation where SJ is on the left of K. How can the nonterminal K appear in this derivation? If the production rule that created K created a suffix of the string SJ at the same time, then the membership of SJ in Left(K) will be disclosed according to the rule (2). On the other hand, if K was the first symbol in a string generated by some other nonterminal L, then (because of rule (3)) it is enough to check that SJ belongs to Left(L). It remains to apply the same argument to L and so on.

In terms of an LR-process, the same idea may be expressed as follows. First, the nonterminal K may participate in several reductions that do not touch SJ (those reductions correspond to applications of the rule (3)). Then a reduction that touches SJ is performed (this reduction corresponds to an application of rule (2)).

It remains to determine which situations are coherent with the empty string; that is, for which nonterminals K, the empty string belongs to Left(K). This can be done according to the following rules:

(1) the initial nonterminal (the axiom) has this property;
(2) if K has this property, K → V is a production rule, and the string V starts with a nonterminal L, then L has this property as well. □

**16.1.10.** Perform the above analysis on the grammar

$$E \rightarrow E + T$$
$$E \rightarrow T$$
$$T \rightarrow T * F$$
$$T \rightarrow F$$
$$F \rightarrow x$$
$$F \rightarrow ( E )$$

(which generates the same language as the grammar of Example 3, p. 223).

*Solution.* The sets State(S) for different S are shown in the table, p. 249. The equals sign means that the sets of situations that are values of the function State(S) of the strings connected by the the equals sign are equal.

Here is the rule to find State(SJ), provided we know State(S) and J (here S is a string of terminals and nonterminals, and J is a terminal or nonterminal):

Find State(S) in the right column; consider the corresponding string T in the left column; append the symbol J to the end of T and find the set corresponding to the string TJ. (If the string TJ is not in the table, then State(SJ) is empty.)                                                                    □

## 16.2 LR(0)-grammars

Recall that our goal is to find a derivation for a given string. In other words, we seek a successful LR-process on this string. We always assume that (for a given string) there exists only one successful LR-process on it. We find this process stepwise. At any point, we find the only possible next step. To ensure that only one step is possible, we need to put some requirements on the grammar. In this section we consider the simplest case, the so-called LR(0)-*grammars*.

As we already know:

(1) The reduction according to the rule K → U with stack S may appear in a successful LR-process if and only if S belongs to LeftCont(K → U) or, equivalently, if S is coherent with situation K → U_.

A similar statement about shift is as follows:

(2) A shift with next symbol a and stack S may appear in a successful LR-process if and only if S is coherent with some situation of type K → U_aV.

**16.2.1.** Prove the above claim.

[*Hint.* Assume that a shift occurs and a new terminal a is added to the stack S. Consider the first reduction that includes this terminal.]                            □

Assume that some grammar is fixed. Consider an arbitrary string S of terminals and nonterminals. If the set State(S) contains a situation where the underscore sign

| String S | State(S) |
|---|---|
| empty | E → _E+T   E → _T   T → _T*F  <br> T → _F   F → _x   F → _(E) |
| E | E → E_+T |
| T | E → T_   T → T_*F |
| F | T → F_ |
| x | F → x _ |
| ( | F → (_E)   E → _E+T   E → _T  <br> T → _T*F   T → _F   F → _x   F → _(E) |
| E+ | E → E+_T   T → _T*F   T → _F  <br> F → _x   F → _(E) |
| T* | T → T*_F   F → _x   F → _(E) |
| (E | F → (E_)   E → E_+T |
| (T | = T |
| (F | = F |
| (x | = x |
| (( | = ( |
| E+T | E → E+T_   T → T_*F |
| E+F | = F |
| E+x | = x |
| E+( | = ( |
| T*F | T → T*F_ |
| T*x | = x |
| T*( | = ( |
| (E) | F → (E)_ |
| (E+ | = E+ |
| E+T* | = T* |

State(S), problem 14.1.10

is followed by a terminal, we say that the string S *allows a shift*. If the set State(S) contains a situation where the underscore sign is the last symbol, we say that the string S *allows a reduction* (according to the corresponding rule). We say that there is a *shift/reduce* conflict for the string S if both shift and reduction are allowed. We say that there is a *reduce/reduce* conflict for S if the string S allows a reduction according to two different rules.

The grammar is called an LR(0)-*grammar* if it has no conflicts of type shift/reduce and reduce/reduce for any string S.

**16.2.2.** Is the grammar given above (with nonterminals E and T) an LR(0)-grammar?

*Solution.* No, it has shift/reduce conflicts for strings T and E+T.                    □

**16.2.3.** Are the following grammars LR(0)-grammars?

$$
\begin{array}{ll}
\text{(a) } T \to 0 & \text{(b) } T \to 0 \\
\quad T \to T1 & \quad T \to 1T \\
\quad T \to TT2 & \quad T \to 2TT \\
\quad T \to TTT3 & \quad T \to 3TTT
\end{array}
$$

*Solution.* Yes, see the corresponding tables (a) and (b) (no conflicts).            □

| String S | State(S) |
|----------|----------|
| empty string | $T \to \_0 \quad T \to \_T1 \quad T \to \_TT2 \quad T \to \_TTT3$ |
| 0 | $T \to 0\_$ |
| T | $T \to T\_1 \quad T \to T\_T2 \quad T \to T\_TT3$ |
|   | $T \to \_0 \quad T \to \_T1 \quad T \to \_TT2 \quad T \to \_TTT3$ |
| T1 | $T \to T1\_$ |
| TT | $T \to TT\_2 \quad T \to TT\_T3$ |
|   | $T \to T\_1 \quad T \to T\_T2 \quad T \to T\_TT3$ |
|   | $T \to \_0 \quad T \to \_T1 \quad T \to \_TT2 \quad T \to \_TTT3$ |
| TT2 | $T \to TT2\_$ |
| TTT | $T \to TTT\_3 \quad T \to TT\_2 \quad T \to TT\_T3$ |
|   | $T \to T\_1 \quad T \to T\_T2 \quad T \to T\_TT3$ |
|   | $T \to \_0 \quad T \to \_T1 \quad T \to \_TT2 \quad T \to \_TTT3$ |
| TT0 | $= 0$ |
| TTT3 | $T \to TTT3\_$ |
| TTT2 | $= TT2$ |
| TTTT | $= TTT$ |
| TTT0 | $= 0$ |

(14.2.3, a)

This problem shows that LR(0)-grammars may be left-recursive as well as right-recursive.

**16.2.4.** Assume that an LR(0)-grammar is given. Prove that each string has at most one rightmost derivation. Give an algorithm that checks whether the input string is derivable.

*Solution.* Assume that an arbitrary input string is given. We construct an LR-process on that string stepwise. Assume that the current stack of the LR-process is S. We have to decide whether a shift or reduce action is needed (and which rule should be used in the reduction case). The definition of LR(0)-grammar guarantees that only one action is possible, and all the information needed to make the decision

| String S | State(S) |
|---|---|
| empty string | T → _0   T → _1T   T → _2TT   T → _3TTT |
| 0 | T → 0_ |
| 1 | T → 1_T |
|  | T → _0   T → _1T   T → _2TT   T → _3TTT |
| 2 | T → 2_TT |
|  | T → _0   T → _1T   T → _2TT   T → _3TTT |
| 3 | T → 3_TTT |
|  | T → _0   T → _1T   T → _2TT   T → _3TTT |
| 1T | T → 1T_ |
| 10 | = 0 |
| 11 | = 1 |
| 12 | = 2 |
| 13 | = 3 |
| 2T | T → 2T_T |
|  | T → _0   T → _1T   T → _2TT   T → _3TTT |
| 20 | = 0 |
| 21 | = 1 |
| 22 | = 2 |
| 23 | = 3 |
| 3T | T → 3T_TT |
|  | T → _0   T → _1T   T → _2TT   T → _3TTT |
| 30 | = 0 |
| 31 | = 1 |
| 32 | = 2 |
| 33 | = 3 |
| 2TT | T → 2TT_ |
| 2T0 | = 0 |
| 2T1 | = 1 |
| 2T2 | = 2 |
| 2T3 | = 3 |
| 3TT | T → 3TT_T |
|  | T → _0   T → _1T   T → _2TT   T → _3TTT |
| 3T0 | = 0 |
| 3T1 | = 1 |
| 3T2 | = 2 |
| 3T3 | = 3 |
| 3TTT | T → 3TTT_ |
| 3TT0 | = 0 |
| 3TT1 | = 1 |
| 3TT2 | = 2 |
| 3TT3 | = 3 |

(14.2.3, b)

is contained in State(S). Therefore, we can find the (only possible) next step of the LR-process.                                                                                      □

**16.2.5.** What happens if the input string has no derivation in the grammar?

*Answer.* There are two possibilities: (1) neither a shift nor a reduce action will be possible at some point; (2) all possible shifts have the next symbol different from the actual one.                                                                                □

*Remarks.* 1. When implementing this algorithm, there is no need to compute the set State(S) from scratch for each value of S. These sets may be kept in a stack. (At any point we keep on the stack the sets State(T) for all prefixes T of the current value of S.)

2. In fact, the string S itself is not used at all. It is enough to keep the sets State(T) for all its prefixes T (including S itself).

The algorithm that checks whether a given string is derivable in an LR(0)-grammar uses only some of the information available. Indeed, for each state it knows in advance which action (shift or reduction — and which reduction) is the only possible one. More elaborate algorithms can make a choice using the next input symbol as well as the stack content. Looking at the set State(S), it is easy to say for which input symbols a shift is possible. (It is possible for all terminals that follow the underscore in situations coherent with S.) The more difficult problem is: How do we use the next input symbol to decide if a reduction is possible?

There are two methods: the first is simpler, the second is more powerful. The grammars for which the first method is applicable are called SLR(1)-grammars (S for Simple). The second method uses all available information; these grammars are called LR(1)-grammars. (There is also an intermediate class of grammars called LALR(1)-grammars, discussed below.)

## 16.3 SLR(1)-grammars

Recall that for any nonterminal K we have defined (see p. 238) the set Follow(K). This set consists of terminals that may follow K in strings that are derivable from the initial nonterminal. This set also includes the symbol EOI if K may appear at the end of a derivable string.

**16.3.1.** Prove that if at some point of the LR-process the last symbol of the stack S is K and the process can be finished successfully, then Next ∈ Follow(K).

*Solution.* This fact is an immediate consequence of the definition (recall the correspondence between rightmost derivations and successful LR-processes).      □

Assume that some grammar is fixed. Consider a string S of terminals and nonterminals, and a terminal x. If the set State(S) contains a situation where the underscore is followed by a terminal x, we say that the pair ⟨S, x⟩ *allows a shift*. If the set State(S) contains a situation K → U_ where x ∈ Follow(K), we say that the pair ⟨S, x⟩ SLR(1)-*allows a reduction* (according to the rule K → U). We say that for the

pair ⟨S, x⟩ there is an SLR(1)-*conflict of type shift/reduce*, if both shift and reduction are allowed. We say that for the pair ⟨S, x⟩ there is an SLR(1)-*conflict of type
reduce/reduce* if reductions according to different rules are allowed.

The grammar is called an SLR(1)-*grammar* if it has no SLR(1)-conflicts of type
shift/reduce and reduce/reduce for all pairs ⟨S, x⟩.

**16.3.2.** Assume that an SLR(1)-grammar is given. Prove that each string has at
most one rightmost derivation. Give an algorithm to check whether a given string is
derivable in the grammar.

*Solution.* We repeat the argument used for LR(0)-grammars. The difference is
that the choice of the next action depends on the next input symbol (Next).      □

**16.3.3.** Check if the grammar shown above on p. 248 (having nonterminals E, T
and F) is an SLR(1)-grammar.

*Solution.* Yes; both conflicts that prevent it from being an LR(0)-grammar are
resolved when we take the next input symbol into account. Indeed, for both T and
E+T a shift is possible only when Next = *, and the symbol * belongs neither to
Follow(E) = {EOI, +, )} nor to Follow(T) = {EOI, +, *, )}. Therefore, reduction is
impossible when Next = *.                                                         □

## 16.4 LR(1)-grammars, LALR(1)-grammars

The SLR(1) approach still does not use all available information to decide if reduction is possible. It checks (for a given rule) whether reduction is possible with a given
stack content and *separately* checks whether reduction is possible with a given input symbol Next. However, these tests are not independent. It may happen that both
checks give a positive answer, but nevertheless the reduction for the given S *and* the
given Next is impossible. The LR(1)-approach is free of this deficiency.

The LR(1)-approach is as follows: All our definitions and statements are modified to take into account what symbol is on the right of the replaced nonterminal
while using a production rule. In other words, we carefully inspect the next symbol
when reduction is performed.

Let K → U be one of the production rules, and let t be a terminal or a special
symbol EOI (which is assumed to be at the end of the input string). We define the
set LeftCont(K → U, t) as the set of all strings that may be the stack content immediately before the reduction U to K during a successful LR-process, assuming that
Next = t at the time of reduction.

All strings in LeftCont(K → U) have suffix U. If we discard this suffix, we obtain
the set of all strings that appear in the rightmost derivations immediately before the
nonterminal K followed by t. This set (which does not depend on the specific rule
K → U, but only on the nonterminal K) is denoted by Left(K, t).

**16.4.1.** Write a grammar whose nonterminals generate the sets Left(K, t) for all
nonterminals K of the given grammar.

*Solution.* Nonterminals are symbols ⟨LeftK t⟩ for any nonterminal K and any terminal t (and also for t = EOI). Its production rules are as follows: Let P be the initial nonterminal (the axiom) of the given grammar. Then our new grammar has the rule

⟨LeftP EOI⟩ →        (the right-hand side is the empty string).

Each rule of the given grammar produces several rules of the new one. For example, if the given grammar has a rule

$$K \to L\,u\,M\,N$$

(L, M, N are nonterminals, u is a terminal), then the new grammar has rules

$$\langle \text{LeftL } u\rangle \to \langle \text{LeftK } x\rangle$$

for all terminals x;

$$\langle \text{LeftM } s\rangle \to \langle \text{LeftK } y\rangle\,L\,u$$

for any s that may appear as a first character in a string derivable from N, and for any y, as well as for all pairs s = y, if the empty string is derivable from N); and

$$\langle \text{LeftN } s\rangle \to \langle \text{LeftK } s\rangle\,L\,u\,M$$

for any terminal s.                                                                      □

**16.4.2.** How should we modify the definition of a situation?

*Solution.* Now a situation is defined as a pair

[situation in the old sense, terminal or EOI]                              □

**16.4.3.** How to modify the definition of a string coherent with a situation?

*Solution.* The string S (of terminals and nonterminals) is coherent with the situation [K → U_V, t] (here t is a terminal or EOI) if U is a suffix of S; that is, if S = TU, and, moreover, T belongs to Left(K, t).                                                □

**16.4.4.** Show how to compute inductively the set State(S) of all situations coherent with a given string S.

*Answer.*

(1) If a string S is coherent with a situation [K → U_V, t] and the first character in V is J; that is, V = JW, then SJ is coherent with the situation [K → UJ_W, t].

This rule determines completely which situations that do not start with underscore are coherent with SJ. It remains to find out for which nonterminals K and terminals t the string SJ belongs to Left(K, t). This is done according to the following two rules:

(2) If the situation [L → U_V, t] is coherent with SJ (according to (1)) and V starts with a nonterminal K, then SJ belongs to Left(K, s) for any terminal s that may appear as a first symbol in a string derivable from V \ K (the string V without the first symbol K), as well as for s = t, if the empty string is derivable from V \ K.

(3) If SJ is in Left(L, t) for some L and t, and L → V is a production rule, and V starts with a nonterminal K, then SJ belongs to Left(K, s) for any nonterminal s that may appear as a first symbol in a string derivable from V \ K, as well as for s = t, if the empty string is derivable from V \ K. □

**16.4.5.** Give the definition of the shift/reduce and shift/shift conflicts in the LR(1)-case.

*Solution.* Assume that a grammar is fixed. Let S be an arbitrary string of terminals and nonterminals. If the set State(S) contains a situation where the underscore sign is followed by a terminal t, we say that the pair ⟨S, t⟩ *allows a shift*. (This definition is the same as in the SLR(1)-case; we ignore the second components of pairs in State(S).)

If State(S) contains a situation whose first component ends with the underscore sign and the second component is a terminal t, we say that the pair ⟨S, t⟩ LR(1)-*allows a reduction* (via the corresponding rule). We say that there is an LR(1)-*conflict of type shift/reduce* for a pair ⟨S, t⟩ if this pair allows both shift and reduction. We say that there is an LR(1)-*conflict of type reduce/reduce* for a pair ⟨S, t⟩ if this pair allows reductions according to different rules. □

The grammar is called an LR(1)-*grammar*, if there are no LR(1)-conflicts of type shift/reduce and reduce/reduce for all pairs ⟨S, t⟩.

**16.4.6.** For any LR(1)-grammar, construct an algorithm that checks if a given string is derivable in the grammar.

*Solution.* As before, at each stage of the LR-process we can determine which action is the only possible one. □

It is useful to understand how the notions of LR(0)-coherence and LR(1)-coherence are related. (It is used below, when LALR(1)-grammars are considered.)

**16.4.7.** Find and prove the connection between the notions of LR(0)-coherence and LR(1)-coherence.

*Solution.* Assume that a grammar is fixed. The string S of terminals and nonterminals is LR(0)-coherent with situation K → U_V if and only if it is LR(1)-coherent with the pair [K → U_V, t] for some terminal t (or for t = EOI). In other words, Left(K) is the union of the sets Left(K, t) for all t. (In the latter form, the statement is almost obvious.) □

*Remark.* Thus the function State(S) in the LR(1)-sense is an extension of the function State(S) in the LR(0)-sense: State$_{LR(0)}$(S) is obtained from State$_{LR(1)}$(S) when we discard the second components of all pairs.

We now give a definition of an LALR(1)-grammar. Assume that a context-free grammar is fixed, S is a string of terminals and nonterminals, and t is a terminal (or EOI). We say that the pair $\langle S, t \rangle$ LALR(1)-*allows a reduction* (according to some production rule) if there is another string $S_1$ with $\text{State}_{LR(0)}(S_0) = \text{State}_{LR(0)}(S_1)$ such that the pair $\langle S_1, t \rangle$ LR(1)-allows reduction according to that rule. Thereafter, the conflicts are defined in a natural way and a grammar is called an LALR(1)-*grammar* if there are no conflicts.

**16.4.8.** Prove that every SLR(1)-grammar is an LALR(1)-grammar and every LALR(1)-grammar is an LR(1)-grammar.

[*Hint.* This is an easy consequence of the definitions.]                       □

**16.4.9.** Find an algorithm that checks if an input string is derivable in an LALR(1)-grammar. This algorithm should keep less information in the stack than the corresponding LR(1)-algorithm.

[*Hint.* It is sufficient to store the sets $\text{State}_{LR(0)}(S)$ in the stack, because the LALR(1)-possibility of reduction is determined by those sets. (Therefore, the only difference with the SLR(1)-algorithm is in the table of possible reductions.)]       □

**16.4.10.**   Construct an LALR(1)-grammar that is not an SLR(1)-grammar.       □

**16.4.11.**   Construct an LR(1)-grammar that is not an LALR(1)-grammar.       □

## 16.5 General remarks about parsing algorithms

Practical applications of the methods described is a delicate matter. (For example, we need to store the tables as compactly as possible.) Sometimes a given grammar is not an LL(1)-grammar but still is an LR(1)-grammar. Often the given grammar can be transformed into an equivalent LL(1)-grammar. It is not clear which of these two approaches is more practical. The following general rule may be useful. If you design the language, keep it simple and do not use the same symbols for different purposes. Then usually it is easy to write an LL(1)-grammar or a recursive-descent parser. However, if the language is already defined by an LR(1)-grammar that is not LL(1), it is better not to change the grammar, just write an LR(1)-parser. To do this, you may use tools for automatic parser generation such as yacc (UNIX) and bison (GNU).

A lot of useful information about the theoretical and practical aspects of parsing can be found in the well-written book of Alfred V. Aho, Ravi Sethi, and Jeffrey D. Ullman on compilers [2].

# Further reading

1. A.V. Aho, J.E. Hopcroft, J.D. Ullman, *The Design and Analysis of Computer Algorithms*, Reading, MA, Addison-Wesley, 1976.

2. A.V. Aho, R. Sethi, J.D. Ullman. *Compilers: Principles, Techniques and Tools*. Reading, MA, Addison-Wesley, 1986.

3. T.H. Cormen, C.E. Leiserson, R.L. Rivest, C. Stein, *Introduction to Algorithms*. Cambridge (Mass.), MIT Press. Third edition, 2009.

4. S. Dasgupta, C.H. Papadimitriou, U.V. Vazirani, *Algorithms*. McGraw-Hill, 2006.

5. E.W. Dijkstra. *A discipline of programming*. Englewood Cliffs, NJ, Prentice Hall, 1976.

6. M.R. Garey, D.S. Johnson. *Computers and Intractability: A Guide to the Theory of NP-completeness*. San Francisco, Freeman, 1979.

7. D. Gries. *The Science of Programming*. New York, Springer, 1981.

8. B. Kernighan, D. Ritchie, *The C language*, Prentice-Hall, 1978; 2 ed., 1988.

9. A.G. Kushnirenko, G.V. Lebedev. *Programming for mathematicians* (*Programmirovanie dlja matematikov*). Moscow, Nauka, 1988.

10. W. Lipski. *Kombinatoryka dla programistów*. Warszawa, Wydawnictwa naukowo-techniczne, 1982. 3rd ed., 2004

11. E.M. Reingold, J. Nievergelt, N. Deo. *Combinatorial Algorithms. Theory and Practice*. Englewood Cliffs, NJ, Prentice Hall, 1977.

12. M. Sipser. *Introduction to the Theory of Computation.* Boston, PWS Publishing Company, 1996.

13. N. Wirth. *Systematic Programming: An Introduction.* Englewood Cliffs, NJ, Prentice-Hall, 1973.

14. N. Wirth. *Algorithms + Data Structures = Programs.* Englewood Cliffs, NJ, Prentice-Hall, 1976.

# Appendix: C and Pascal examples

We used Pascal notation in this book. However, it is quite easy to convert our program examples into any other procedural language. In this appendix we give examples of such conversions to C, another classical language.

The most trivial (though annoying) difference is the usage of the equality sign. While C uses it for assignment statements, Pascal uses := for assignments, reserving = for equality relation; C uses == instead. There are some other differences in the notation for relations: $\neq$ is denoted by <> in Pascal and by != in C. The notation for $<, \leqslant, >$ and $\geqslant$ is the same (<, <=, > and >=). The logical operations (and, or and not in Pascal) are denoted by &&, || and ! in C. The operations div and mod are denoted in C by / and % (assuming that the operands are of integer type; if one of the operands has real type, / means division). Fortunately, +, -, * have the same meaning in Pascal and C. Other differences: Instead of special Boolean type the integer type is used in C; constants True and False are represented as 1 and 0. Finally, the comments are put in /* ... */ brackets instead of { ... } or (* ... *) in Pascal.

| PASCAL | C |
|---|---|
| a := b; | a = b; |
| u := True; | u = 1; |
| a = b | a == b |
| a <> b | a != b |
| (X and Y) or not Z | (X && Y) \|\| !Z |
| a mod b | a % b |
| a div b | a / b |
| { comment } | /* comment */ |

The declarations of variables also are slightly different; the most annoying difference is that in C array elements are always numbered starting from 0. So only the size of the array needs to be declared. The array elements are denoted still in the same way. Note also that `real` type in Pascal is called `float` in C; one can also use `double` which may have better precision (and occupy more memory).

```
PASCAL                              C

var a,b: integer;                   int a,b;

 x: real;                           float x;

 c: char;                           char c;

 s: array [0..255] of char;         char s[256];

 t: array[0..15,0..7] of char;      char t[16,8];

 t[a][b] = s[b]                     t[a][b] == s[b]
```

The control structures also have some differences: `if–then` construction is represented as follows:

```
if a=b then begin                   if (a==b) {
    <A>;                                <A>;
    <B>;                                <B>;
end;                                }
```

The conditional statement with `else`-part:

```
if a=b then begin                   if (a==b) {
    <A>;                                <A>;
    <B>;                                <B>;
end else begin                      }else{
    <C>;                                <C>;
end;                                }
```

The `while`-construction is represented as follows:

```
while a=b do begin                  while (a==b) {
    <A>;                                <A>;
    <B>;                                <B>;
end;                                }
```

A more subtle difference occurs in the procedures and function declarations and the parameter specification. Things are almost the same if the function does not change its parameters; the assignment to the function name is replaced by the `return` construction.

| PASCAL | C |
|---|---|

```
function m(a,b:integer):integer;      int m (int a,b)
begin                                 {
  if (a>b) then begin                   if (a>b){
    m:=a;                                 return(a);
  end else begin                        }else{
    m:=b;                                 return(b);
  end;                                  }
end;                                  }
```

The C language does not have procedures but uses function declarations with special return type `void`. However, to be useful, they need to change the parameters. This is not allowed in C. Instead, one can use the *address* (memory location) of some variable as a parameter. The declaration

```
int *a;
```

says that a is a variable whose value is the address of some memory location suitable for keeping an integer value. Then the assignment (say)

```
*a = 1917;
```

puts the integer value 1917 into this location (but does not change the value of the variable a that remains the address of the same memory location). If b is an integer variable (properly declared), then &b is the address of this variable, so the assignments b=1917 and *(&b)=1917 both change the value of b. This notation can be illustrated by a procedure that exchanges the values of two integer variables:

```
procedure swap(var a,b:integer);     void swap (int *a,*b)
begin                                 {
  var tmp: integer;                     int tmp;
  tmp := a;                             tmp = *a;
  a := b;                               *a = *b;
  b := tmp;                             *b = tmp;
end;                                  }
```

The call of this procedure (that exchanges the values of two integer variables x and y) then looks like:

```
swap (x,y);                          swap (&x,&y);
```

C replaces the type declarations by a `typedef` construction and uses the so-called "structures" instead of "records"; however, here the differences are quite significant and since we rarely use these constructs in our examples, we do not describe them here. (See the classical textbook written by the inventors of C language, B. Kernighan and D. Ritchie, *The C language*, Prentice-Hall, 1978; 2 ed., 1988.)

We conclude this appendix by translating one of the programs in Chapter 3 (page 56; we omit the comments) from Pascal to C. Both original Pascal program and its C translation are ready-to-compile versions.

The Pascal program starts with name, constants and global variables:

```
program queens;
  const n = 8;
  var
    k: 0..n;
    c: array [1..n] of 1..n;
```

In C the program does not have a name. Instead of `const` we use `#define` construction that means that N should be replaced everywhere by 8. The array c is one element bigger in the C version since we want to keep the numbering starting from 1 (the element c[0] remains unused). The line `#include <stdio.h>` says that definitions from a standard header file `stdio.h` should be used; the same for `stdlib.h`. The bounds 0..n and 1..n are not used, values are just integers.

```
#include <stdio.h>
#include <stdlib.h>
#define N 8
int k;
int c[N+1]; /* c[0] unused */
```

Function definitions:

```
procedure begin_work;
begin
  k := 0;
end;
```

is translated as

```
void begin_work(){
  k=0;
}
```

(`void` is used because the procedure does not return any value; we use `()` to show that it does not have parameters). Next function:

```
function danger: Boolean;
  var b: Boolean; i: integer;
begin
  if k <= 1 then begin
    danger := false;
```

```pascal
      end else begin
        b := false; i := 1;
        while i <> k do begin
          b := b or (c[i]=c[k])
                or (abs(c[i]-c[k])=abs(i-k));
          i := i+1;
        end;
        danger := b;
      end;
  end;
```

In C, Boolean variables are represented by integers ($0 = $ false, $1 = $ true) and
return is used to specify the returned value. The statement i++; is a standard C
idiom for increasing the value of a variable i by 1.

```c
      int danger(){
        int b,i;
        if (k<=1){
          return(0);
        }else{
          b=0; i=1;
          while (i!=k){
            b=b||(c[i]==c[k])||(abs(c[i]-c[k])==abs(i-k));
            i++;
          }
          return(b);
        }
      }
```

Next three functions

```pascal
      function is_up: Boolean;
      begin
        is_up := (k < n) and not danger;
      end;

      function is_right: Boolean;
      begin
        is_right := (k > 0) and (c[k] < n);
      end;

      function is_down: Boolean;
      begin
        is_down := (k > 0);
      end;
```

are translated in a straightforward way:

```
int is_up(){
  return((k<N) && !danger());
}

int is_right(){
  return ((k>0) && (c[k]<N));
}

int is_down(){
  return (k>0);
}
```

The procedures for tree traversal,

```
procedure up_left;
begin
  k := k + 1;
  c [k] := 1;
end;

procedure right;
begin
  c [k] := c [k] + 1;
end;

procedure down;
begin
  k := k - 1;
end;
```

again become void functions; we use increment (++) and decrement (−−) notation:

```
void up_left(){
  k++;
  c[k]=1;
}

void right(){
  c[k]++;
}

void down(){
  k--;
}
```

Translating the next procedure

```
procedure process;
```

```
      var i: integer;
    begin
      if (k = n) and not danger then begin
        for i := 1 to n do begin
          write ('<', i, ',' , c[i], '> ');
        end;
        writeln;
      end;
    end;
```

we use two constructions that are not yet explained. First,

$$for(A;B;C)\{D\}$$

means

$$A; \ while(B)\{D;C\}$$

and is useful to replace for-loop in Pascal. Then function printf is used. Its first
argument is a string that should be printed after %d-templates are replaced by decimal
representations of the other arguments that are assumed to be integer values. (There
are many other types of templates that we do not need in this program.) Finally, \n
stands for the newline character produced by writeln in Pascal.

```
    void process(){
      int i;
      if ((k==N) && !danger()){
        for (i=1;i<=N;i++){
          printf("<%d,%d> ", i, c[i]);
        }
        printf("\n");
      }
    }
```

The next procedure,

```
    procedure go_up_and_process;
    begin
      while is_up do begin
        up_left;
      end;
      process;
    end;
```

is copied almost literally:

```
    void go_up_and_process() {
      while (is_up()){
        up_left();
      }
```

```
        process();
    }
```

Then the body of Pascal program,

```
    begin
      begin_work;
      go_up_and_process;
      while is_down do begin
        if is_right then begin
          right;
          go_up_and_process;
        end else begin
          down;
        end;
      end;
    end.
```

is converted into main function (that is called when the C program is executed); traditionally this function is considered as returning an integer value (defined by exit statement) and 0 means successful termination.

```
    int main(){
      begin_work();
      go_up_and_process();
      while (is_down()){
        if (is_right()){
          right();
          go_up_and_process();
        }else{
          down();
        }
      }
      exit(0);
    }
```

# Index

Printed in the United States
By Bookmasters